T0353444

An Exploration into China's Economic Development and Electricity Demand by the Year 2050

An Exploration into China's Economic Development and Electricity Demand by the Year 2050

By

Zhaoguang Hu
Xiandong Tan
Zhaoyuan Xu et al.

AMSTERDAM • BOSTON • HEIDELBERG • LONDON • NEW YORK • OXFORD
PARIS • SAN DIEGO • SAN FRANCISCO • SINGAPORE • SYDNEY • TOKYO

Elsevier
32 Jamestown Road, London NW1 7BY
225 Wyman Street, Waltham, MA 02451, USA

Notice
Knowledge and best practice in this field are constantly changing. As new research and experience broaden our understanding, changes in research methods, professional practices, or medical treatment may become necessary.

Practitioners and researchers must always rely on their own experience and knowledge in evaluating and using any information, methods, compounds, or experiments described herein. In using such information or methods they should be mindful of their own safety and the safety of others, including parties for whom they have a professional responsibility.

To the fullest extent of the law, neither the Publisher nor the authors, contributors, or editors, assume any liability for any injury and/or damage to persons or property as a matter of products liability, negligence or otherwise, or from any use or operation of any methods, products, instructions, or ideas contained in the material herein.

British Library Cataloguing-in-Publication Data
A catalogue record for this book is available from the British Library

Library of Congress Cataloging-in-Publication Data
A catalog record for this book is available from the Library of Congress

ISBN: 978-0-12-420159-0

For information on all Elsevier publications
visit our website at store.elsevier.com

This book has been manufactured using Print On Demand technology. Each copy is produced to order and is limited to black ink. The online version of this book will show color figures where appropriate.

Contents

Preface

China's successful economic and social development following the economic reform has received worldwide attention. We have become increasingly important in the international community, and our living conditions have continuously improved. In 2010, China's GDP and energy consumption both ranked second in the world, with many types of production ranking first. The next 10–20 years will be an important period of strategic opportunities for China's economic and social development. They will also prove to be a crucial period for the construction of a prosperous society and the rejuvenation of the nation. Patterns of economic development will significantly change, economic structure will be constantly optimized, industrialization will be fundamentally implemented, urbanization will reach comparatively high levels, and regional economies will experience profound changes. Further into the future, to the mid-twenty-first century, China's "Three-Step" strategic objectives will be basically completed. The level of electricity demand and the structure of electricity consumption at that time are very basic questions for the country. However, there are lots of challenges on energy, environment, and sustainable development. This book is an exploratory study of China's economic development and electricity demands from 2030 to 2050 and attempts to address the aforementioned challenges.

In 2009, the State Grid Energy Research Institute (SGERI) and the Department of Development Strategy and Regional Economy of DRC began cooperating to develop "A Study of China's Economic Development in the Year 2030," headed by well-known scholar, Professor Shantong Li. "A Study of China's Economic Development in the Year 2030" primarily analyzes China's economic development since the economic reform of the 1980s, casting light on important factors and changing trends in China's economic development. Furthermore, the study uses DRC-CGE modeling to simulate regional economic development across China. The research has been an excellent foundation for this book. This book examines three types of economic development scenarios, encompassing the years 2010–2030, based on Li's research. Using the results of economic development simulations between 2010 and 2030 as the premise, we have gone one step further with the use of Intelligent Laboratory of Economy–Energy–Electricity–Environment (ILE4) to simulate national and regional electricity demand. Furthermore, we also look ahead to the economic development and power demands of 2040 and 2050.

This book has seven chapters: Chapter 1 presents exploration and discovery, gauging the connection between electricity demand and economic growth, and concluding with the main characteristics of China's economic development and electricity consumption since economic reform. In addition, the chapter introduces simulation results for China's economic development and growth of power demands into the year 2050. Finally, an international comparison is made based on prior research, and 10 discoveries are extracted resulting in an overview of economic development and growth in electricity consumption. Chapter 2 is an introduction to fundamental principles and basic functions and models used in ILE4, highlighting the laboratory's analytical methods for researching long-term economic development and power demand. Chapter 3 is a review of China's economic development and electricity consumption, summarizing changes since economic reform. Chapter 4 covers principle influences on China's electricity demand, focusing on trends in global economic growth, domestic economic restructuring, changes in economic geography, technological progress, and related effects on electricity consumption. Chapters 5 and 6 contain an analysis and summary of China's economic development and electricity demand in the year 2030. Chapter 7 examines potential economic development and electricity demand in 2050 simulating economic development and electricity demand.

Following are a list of the contributors involved in each chapter: Chapter 1 was written by Zhaoguang Hu; Chapter 2 was written by Zhaoguang Hu, Baoguo Shan, Minjie Xu, Xinyang Han, Quan Wen, and Xiandong Tan; Chapter 3 was written by Yugui Gu, Qing Huang, Baoguo Shan, Lijie Guo, Zheng Si, Zhi Luo, and Xiandong Tan; Chapter 4 was written by Qing Huang, Baoguo Shan, Xiandong Tan, Lijie Guo, Jing Zhao, Peng Wu, Lei Chen, Lu Xing, Zheng Si, Zhi Luo, and Chengjie Wang; Chapter 5 was written by Shantong Li, ZhaoYuan Xu, Baoguo Shan, and Xiandong Tan; Chapter 6 was written by Xiandong Tan, Zhaoguang Hu, Minjie Xu, Baoguo Shan, Xinyang Han, and Quan Wen; Chapter 7 was written by Xiandong Tan, Zhaoguang Hu, and Baoguo Shan. The entire book was overseen by Zhaoguang Hu.

During the compilation of this book, Huijiong Wang, Jifa Gu, Chengzhang Zhu, Ying Ran, Xinmao Wang, Shaojun Jiang, Xian Zhou, Yongsheng Xu, Weiping Hao, Haiping Xiang, Changyu Ouyang, Weiping Huang, Bo Liang, Yuzhi Ren, Yinong Zhao, Zhengling Zhang, Jianfang Zhou, and other experts and scholars gave many constructive suggestions related to topics in this book. From the DRC: Professor Shantong Li, Professor Yunzhong Liu and assistant researcher Jianwu He, and from Tsinghua University: postdoctoral fellows Ming Liu, Sanmang Wu, and Shaojun Zhang, all gave a great deal of assistance for which I would like to express my heartfelt gratitude. In addition, I obtained guidance and assistance from the SGERI Yunzhou Zhang, Xuehao Yu, Zhongbao Niu, Liping Jiang, Ying Li, Xubo Ge and others, for which I am grateful.

It is inevitable that this book contains some omissions and deficiencies, and we are expecting the readers to discuss and criticize the content. Truth develops out of criticism, while falsehood is bred through praise.

The Author
April 12, 2011

Foreword

I was invited by European Countries to participate in a large international conference "Euro prospective" held in Paris on April 1987. This conference was jointly organized by the Centre National de la Recherche Scientifique/CNRS, the Centre for Long-term Forecasting and Evaluation/CPE, the French Planning Commission/CGP, and the Commission of the European Communities/CEC. One of the background of this conference which I felt personally was the high economic growth of Japanese economy during that period and the sense of crisis of weak competitiveness of European Countries. Although it was stated "The objective of Euro prospective is to define the active ferment in Europe and to disseminate visions of the future in their most concrete aspects." The speech I have given in the opening session was "China by the year 2000" which was an organized research project done by our Center. On the day before the conference, Mr. Thierry Gaudin, the director of a French Bureau raised a question to me, "Have you studied I-Ching?" His question impressed me the influence of traditional Chinese culture to abroad. Later on I knew that there are many traditional Chinese studies published in France and England such as I-Ching, the Art of War, acupuncture, fengshui, Chinese Medicine, etc. Many thoughts and descriptions have been explained with modern perspective. It can be seen that how deep and broad are the traditional Chinese culture. To be a Chinese in contemporary world, we must consider seriously how to absorb advanced culture from both domestically and abroad, innovate on them through inheritance, and promote its exchange globally to be a part of progress of historical process of the mankind.

The study of economic development and projections has been lasted around hundred years since last century, due to the popularization of various economic planning both directive and indicative, there are emergence of various economic models and policy models, forecasting techniques, improvement of economic and social statistics, and various indicator system for measurement of development. Many international organizations such as the World Bank, IMF, OECD, the financial institutions and higher educational institutions are engaging in the analysis and projections of economic development. There are emergence of many new methods, the accuracy of projections has been raised. But no institutions had foreseen the tremendous impact to the global economy caused by twice oil crisis in 1970s of last century. They cannot also forecast in time the global financial and economic crises erupted in later part of 2007. Even some main stream financial experts declared that "There is no way to forecast the eruption of financial crisis." Because there are too many factors which can effect the reality of economic life. The mankind may require to explore continuously with full effort in the process to recognize and understand the dynamic objective world.

In spite of the economic analytic capability mastered by the mankind which cannot deal with several limited special issues, but the current methods of economic analysis and demand forecasting are still the necessary tools in determining the development strategy, objectives, and measures of countries in the world in most cases. Vice President Zhaoguang Hu and his team, researchers at the State Grid Energy Institute, had been involved to explore the relationship between economic development and electricity demand for quite a long period. This publication *Exploration of China's Economic Development and Electricity Demand: 2050* edited by him and Mr. Xiandong Tan and Mr. Zhaoyuan Xu is an important reference for China's long- and medium-term economic development, it is also a contribution to the planning and layout of generation and transmission capacity of electric system. I do think that there are many features of the values and meanings of this publication. Only three aspects will be described simply below:

1. Chapter 2 contains a brief introduction to Intelligent Laboratory of Economy–Energy–Electricity–Environment (ILE4) which was established in the State Grid Energy Research Institute (SGERI) by the State Grid Corporation of China in 2008. In this chapter, basic theory, principle functions, and principle models of ILE4 are introduced, as well as methods to analyze the economy and electricity demand. Around the 1990s, there were some domestic institutions engaged in the construction of data base, model base, and method base; however, SGERI has a unique comparative advantage, i.e., it owns the daily and monthly instantaneous data of power load and electricity. The Bureau of Economic Analysis of US Department of Commerce created the leading, coincident, and lagging indicators system and initiated the publication of monthly reports to help companies to analyze economic trends and business cycles. However, the amount and change of electrical power and electricity are also one of the best coincident economic indicators from my perspective, although they have not previously been used by economists. ILE4 of SGERI has the privilege to collect information timely and thereby enabling them to make improvements to their economic analysis work and provide accurate comparative analyses. Furthermore, this laboratory has taken the lead to adopt generalized models, intelligent space, and intelligent paths as core methods of intelligent engineering theory. It is an experiment of bold innovation. It is expected that they can summarize experiences in continuous to update and perfect their work.

2. This book will illustrate the broad-based knowledge from various sources of the chief editors and the efforts of their team work. It is also the result of well-organized research. Along with the rapid development and continuous progress of contemporary society, global society is marching from industrialized society to post-industrialized society, or knowledge society. Due to imbalances of global development, a small number of societies are still in the stage of agricultural or even hunting society. However, in the stage of knowledge societies, the accumulated and created knowledge of the mankind is so vast and diversified, that can hardly be mastered by individuals. Therefore, modern research requires extensive cooperation to achieve the result that closed relatively to the reality of objective world. Authors of this book, besides Vice President Zhaoguang Hu and his team of economic engineering researchers at SGERI, also consist of Professor Shantong Li and her team from Development Research Center of the State Council (DRC). The latter has long been engaged in China's development strategies and policy research, as well as mathematical modeling. Close collaboration of these two teams can realize a competitive advantage through complementarity of comparative advantage of each other. This organized work is just what is needed in our current knowledge society and gives the contents of this book a unique and extraordinary feature.

3. In Section 1.3, China's economic cycle has been explored through simulation of growth of GDP and electric power based upon statistics from National Bureau of Statistics and China's Electricity Council. Many useful experiences can be obtained if China's real economic activity can be analyzed linked with the economic cycles obtained from simulation of this book. Due to the limitation of scale of this book, there is not in-depth analysis on this subject. Anyhow, this book has provided very meaningful raw materials.

I do believe that the publication of this book will provide useful knowledge and hint to readers and researchers of public policy. I also expect the people worked in technology and economy will continuously work on the exploration of economic analysis and mathematical economic modeling. This is also a contribution to the cultural exchange of China to other parts of the world. That is the reason I felt honored to write a foreword for this publication.

Huijiong Wang

Development Research Center of the State Council, China

April 20, 2011

Summary

This book reviews the main features of China's economic development and electricity consumption since the economic reform of the 1980's. It includes analyses of the intrinsic connection between electricity demand and economic growth and the changing trends of the adjustment of economic structure, regional layout optimization and development of the energy intensive industry, as well as how these factors impact China's electricity demand. In addition, the book proposes three possible scenarios in the next 20 years of China's economic development and growing demand for electricity based on the detailed simulations conducted by Intelligent Laboratory of Economy-Energy-Electricity-Environment (ILE4) in regional economic development and electricity demand in 2030 as well as the prospective of China's electricity demand and economic growth in the year 2050. By means of this exploratory analysis, we hope to demonstrate the future trends in the trajectory of China's economic development and electricity demand.

This book is particularly suitable for those electric power system planning technicians, stock market analysts, economists, and government policy makers in relevant fields.

List of Contributors

Lei Chen   Department of Economy and Energy Supply & Demand research, State Grid Energy Research Institute, Xicheng, Beijing, China.

Yugui Gu   Department of Economy and Energy Supply & Demand research, State Grid Energy Research Institute, Xicheng, Beijing, China.

Lijie Guo   Department of Economy and Energy Supply & Demand research, State Grid Energy Research Institute, Xicheng, Beijing, China.

Xinyang Han   Department of Economy and Energy Supply & Demand research, State Grid Energy Research Institute, Xicheng, Beijing, China.

Jianwu He   Department of Development Strategy and Regional Economy, Development Research Center of the State Council, Dongcheng, Beijing, China.

Zhaoguang Hu   State Grid Energy Research Institute, Xicheng, Beijing, China.

Qing Huang   Department of Economy and Energy Supply & Demand research, State Grid Energy Research Institute, Xicheng, Beijing, China.

Shantong Li   Department of Development Strategy and Regional Economy, Development Research Center of the State Council, Dongcheng, Beijing, China.

Yunzhong Liu   Department of Development Strategy and Regional Economy, Development Research Center of the State Council, Dongcheng, Beijing, China.

Ming Liu   School of Public Policy and Management, Tsinghua University, Haidian, Beijing, China.

Zhi Luo   Department of Economy and Energy Supply & Demand research, State Grid Energy Research Institute, Xicheng, Beijing, China.

Baoguo Shan   Department of Economy and Energy Supply & Demand research, State Grid Energy Research Institute, Xicheng, Beijing, China.

Zheng Si   Department of Economy and Energy Supply & Demand research, State Grid Energy Research Institute, Xicheng, Beijing, China.

Xiandong Tan   Department of Economy and Energy Supply & Demand research, State Grid Energy Research Institute, Xicheng, Beijing, China.

Chengjie Wang   Department of Economy and Energy Supply & Demand research, State Grid Energy Research Institute, Xicheng, Beijing, China.

Quan Wen   Department of Economy and Energy Supply & Demand research, State Grid Energy Research Institute, Xicheng, Beijing, China.

Peng Wu   Department of Economy and Energy Supply & Demand research, State Grid Energy Research Institute, Xicheng, Beijing, China.

Sanmang Wu   School of Public Policy and Management, Tsinghua University, Haidian, Beijing, China.

Lu Xing   Department of Economy and Energy Supply & Demand research, State Grid Energy Research Institute, Xicheng, Beijing, China.

Minjie Xu   Department of Economy and Energy Supply & Demand research, State Grid Energy Research Institute, Xicheng, Beijing, China.

Zhaoyuan Xu   Enterprise Research Institute, Development Research Center of the State Council, Dongcheng, Beijing, China.

Shaojun Zhang   School of Public Policy and Management, Tsinghua University, Haidian, Beijing, China.

Jin Zhao   Department of Economy and Energy Supply & Demand research, State Grid Energy Research Institute, Xicheng, Beijing, China.

Fagen Zhu   Department of Economy and Energy Supply & Demand research, State Grid Energy Research Institute, Xicheng, Beijing, China.

1 Exploration and Discovery

1.1 Discovery

China's economy has developed rapidly in the thirty plus years since the economic reform in 1978, that it is now the second largest economy in the world. In 2010, GDP reached 31.4 trillion Yuan (in constant 2005 Yuan),[1] and consumption of electricity was reaching 4.19 trillion kWh.[2] Many people follow future (long-term) trends of Chinese economic development and electricity demand closely and are especially interested in how development will look like in 2030 and 2050. There are numerous factors influencing the economy and electricity demand, and the combination of all these factors would produce astronomical fi gures. How to correctly judge the influence of each of these factors is a global conundrum. In the history of the world, there has never been anyone able to correctly make forecasts about the economy and demand for electricity. However, this has not stopped people from conducting research and exploration in this field. Normally, people can adopt a scenario analysis method, conducting research to provide hypotheses for some uncertain parameters in the future. By using generalized models they can provide depictions of factors that are difficult to measure. Through their own assessment of the combination of these numerous factors, they can choose several relatively high probability cases as scenarios in order to provide a hypothesis, which they can then test through further research.

The results of these forecasts are actually just to outline for people a vision of the future. Accuracy is in fact not important. If during the process of research and exploration, one is able to discover something new, uncover the tiniest of revelations, or dig up something of value, then the forecast becomes much more meaningful. Perhaps the journey of exploration is more worthy of our appreciation than the results.

As for research into China's long-term economic development and electricity demand, a study has been conducted with Intelligent Laboratory of Economy-Energy-Electricity-Environment (ILE4), during which we set three scenarios as detailed simulations for the period between 2010 and 2030. We then conducted trend simulation for economic development and electricity demand from 2030 to 2050 based on the first of the 2010–2030 scenarios. Finally, we analyzed trends of Chinese economic development and electric power demand from 1980 to 2050. Through summarizing past research and conducting international comparisons, we have got 10 discoveries as follows:

1. *Patterns*: In modern economic development, the growth of electricity consumption is usually greater than the growth of primary energy consumption.

An Exploration into China's Economic Development and Electricity Demand by the Year 2050.
DOI: http://dx.doi.org/10.1016/B978-0-12-420159-0.00001-1

2. *Oscilloscope*: There is a close relationship between the electricity and the economy. Changes in electricity consumption correctly, reliably, and dynamically reflect economic trends, and can be seen as a sort of economic oscilloscope.

3. *Properties*: Looking at economics through the lens of electricity, the Chinese economy has a cyclic property of about 9 years. It resembles a sinusoidal wave with rolling development and continuously decreasing amplitude.

4. *Characteristics*: When a country (or region) completes the industrialization process, its electricity consumption per capita is approximately 4500–5000 kWh, and its residential electricity consumption per capita is approximately 900 kWh.

5. *Risk*: It can be seen from comparisons of China's cyclic development that there may be a relatively large risk of recession after large shifts in economic growth (before 2020).

6. *Opportunity*: Before China completes industrialization (2020), tertiary industry has a large space for development. Technology such as electric automobiles will promote economic activity in China during the period between 2020 and 2030, causing the demand for electric power to continue increasing steadily after 2030.

7. *Vision*: In 2030, China's economic output will reach approximately 118 trillion Yuan (in 2005 Yuan), and GDP per capita will reach approximately 80,000 Yuan. Electricity demand will reach about 9.9 trillion kWh, and electricity demand per capita will reach about 6272 kWh. In 2050, economic output will reach 273 trillion Yuan, and GDP per capita will reach approximately 180,000 Yuan. Electricity demand will reach about 14.3 trillion kWh, and electricity demand per capita will reach about 9207 kWh.

8. *Gaps*: Although China's consumption of electricity is among the highest in the world, its electricity consumption per capita and residential electricity consumption per capita are low. Compared to the international level of electricity consumption per capita, there still exists a large gap. China's electricity consumption per capita in 2010 was close to that of the United States in 1955 and will approach the United States' 1980 electricity demand per capita in 2050. China's 2010 level of electricity consumption per capita was similar to Japan's in 1970 and will exceed Japan's 2010 level by 2050. In 2050, China will have a large number of electric automobiles on the road, while the United States did not in 1980, and Japan did not in 2010.

9. *Saturation point*: Without taking into account breakthroughs in electric automobiles and other electricity-intensive technologies but considering constraints such as resources and environment, China's saturation point for electricity demand per capita in future economic and social development will be about 8000 kWh.

10. *Expected breakthrough point*: During the period from 2040 to 2050, China's elasticity of electricity could reach a new global low, breaking through 0.41.

What does it mean that the growth of electric power exceeds energy growth? It means that if the economy is not experiencing zero or negative growth, then the electricity elasticity is larger than the energy elasticity. The electricity elasticity is associated with electricity intensity, while the energy elasticity is associated with energy intensity. It can be proven by mathematics[3]: a decline of electricity intensity if and only if the electricity elasticity is less than 1; a decline of energy intensity if and only if the energy elasticity is less than 1. Similarly, a rise of electricity intensity if and only if the electricity elasticity is greater than 1; an rise of energy intensity if and only if the energy elasticity is greater than 1. According to the rule of electricity elasticity greater than energy elasticity, when the electricity elasticity is less than 1, then it indicates that the decline degree of electricity intensity is smaller than the decline degree of energy intensity. On the contrary, if the energy elasticity is greater than 1, it indicates that the increasing degree of electricity intensity is greater than the increasing degree of energy intensity. When electricity elasticity is greater than 1 and the energy elasticity is less than 1, it indicates that the electricity intensity will increase but the energy intensity will decrease.

Therefore, as long as this rule is followed, and if the greatest effort possible is made to dissolve risks in China's economic development and two opportunities are seized, then the expected breakthrough point will be both hopeful and reachable.

1.2 Positive Correlations Between Electricity Consumption and the Economy

One cannot live or take part in economic activities in modern society without electricity. Electricity is a special type of commodity in the national economy and has two characteristics: The production, transport, and consumption of electricity is completed simultaneously and instantaneously; and any mass storage of electricity is impossible, it requires that a constant and complete balance must be maintained between its generation and consumption. These two characteristics determine the immediacy of electricity production, transmission, and consumption, as well as the synchronization of electricity and the national economy. In addition, electricity data is calculated through the readings of meters at electricity generation side and electricity usage side. As a result, the immediacy, reliability, and accuracy of data on electric power are objectively guaranteed. These two characteristics determine the close correlation between the economic development and the production and consumption of electricity, as well as the vital role of electricity economics in the economic research.

Changes in electricity consumption and the proportion of electricity consumption also reflect the status of the economy and its current stage of development. As the largest economy in the world, from 1949 to 2006, the United States' GDP increased from 1.6346 to 11.4136 trillion USD (in constant 2000 $) with an average annual growth of 3.468%; electricity consumption increased from 0.2545 to 3.8197 trillion kWh with an average annual growth of 4.866%. Plotting the United States' GDP and electricity consumption data on a scatter plot (Figure 1.1), the correlation coefficient between the two is 0.9921, which indicates a positive correlation.

Japan experienced a rapid economy development after World War II, as its GDP increased from 120 (in constant 2000 Yen.) in 1965 to 553.44 trillion Yen in 2006 with an average annual growth of 6.7%. Consumption of electricity rose from 0.16882 to 1.0483 trillion kWh during the same period—an average annual growth of 4.55%. Making a similar scatter plot for Japan's GDP and electricity consumption (Figure 1.2) shows that the correlation coefficient between the two is 0.9903, which is also a positive correlation.

From 1978 to 2009, the correlation coefficient between China's national electricity consumption and GDP (in constant 2000 Yuan) reached 0.993. Figure 1.3 is the

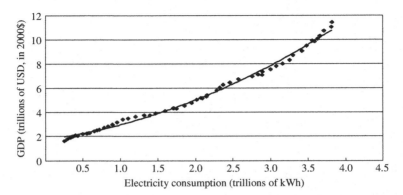

Figure 1.1 Correlation between America's electricity consumption and its GDP.
Source: http://www.census.gov/ and http://www.bea.doc.gov/.

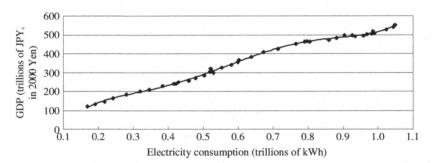

Figure 1.2 Correlation between Japan's electricity consumption and its GDP.
Source: Handbook of energy and economic statistics in Japan, 2008, The Institute of Energy Economics, p. 6, p. 189.

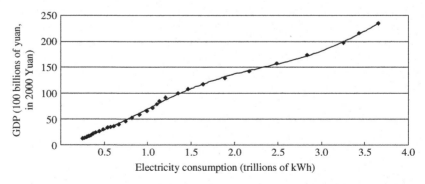

Figure 1.3 Correlation between China's national electricity consumption and its GDP.
Source: CEC Compilation of Statistics on Electric Power, National Bureau of Statistics.

scatter plot of the two. Many more examples likes these can be made to reflect the positive mathematical correlation that exists between electricity consumption and economic output. If GDP reflects a nation's economic strength, then electricity consumption is also one of the material bases that reflect the economic strength.

Magnitude of value is normally used to evaluate the economic level, but value follows the variability of time and the plasticity of currency exchange rates, which makes it difficult to measure. Currency has time value—what one could buy for 100 Yuan in 2000 may cost more in 2010—in other words, it depreciates. Additionally, exchange rates make it relatively difficult to compare exchanges between different currencies. Electricity indices, on the other hand, are physical quantities and are therefore easily measured. They do not change with time, and they do not differ between countries, providing them with strong stability. Because of this, using electric power to discuss stages of industrial progress and trends in economic development in China is not only a supplement to value discussions but also fills a need for physical quantities.

Growth in electricity consumption can be seen as a "thermometer" of economic development, monitoring in real time the economy's "temperature" and recording the trajectory of economic activity. Figure 1.4 shows monthly changes in China's electricity consumption growth in 2008–2009. Before the global economic crisis broke out, many Chinese export companies were affected by a reduction in international orders and production was sluggish. In April of 2008, growth in electricity consumption had already begun to fall. In June, it fell below 10% and stayed at about 6% after that. In September, international finance tycoon, Lehman Brothers, announced its bankruptcy and became the trigger of the financial crisis, which began with this announcement and spread rapidly afterward. Electronic consumption experienced negative growth in October and bottomed out in November at -8.7%. It was at that time that the government quickly made policy decisions to ramp up investment in order to stimulate economic recovery. In June, 2009, after 7 months in negative

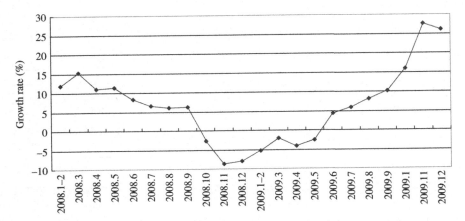

Figure 1.4 2008–2009 Monthly changes in China's national electricity consumption.
Source: CEC Compilation of Statistics on Electric Power.

territory, electricity consumption began to show positive growth, due to the lag effect of investment (about 6 months). In fact, it continued to rise until the end of 2009, when it exceeded 25%. This shows that investment efforts were a bit excessive. In reality, a period of economic downturn is the best opportunity to squeeze out any economic bubbles but the opposite occurred. This was certainly not the intention of the central government, but it shows that under the current system, the relaxing of macro control policies is relatively easy while the tightening of such is rather difficult. Of course, this was also a valuable experience in regard to macro controls—one worth summarizing and studying.

If changes in electricity consumption are used as a "thermometer" for the economy, that would allow for dynamic observation of economic performance. If the economy is found to be "too hot" or "too cold," alarms can be raised and measures can be taken (monetary policy, fiscal policy, etc.) to make timely adjustments, put in place preventative measures, and maintain healthy economic development.

1.3   Characteristics of China's Economic Growth and Electricity Consumption

China's economy has been growing rapidly since 1978, and in 2010 its average annual growth in GDP reached 9.9%, far exceeding initial economic growth in most countries. Research showed that economic growth in China displays a cyclic property of about 9 years.[4] As the simulated GDP curve shown in Figure 1.5, the cycle displays oscillating path resembling a sinusoidal curve, the amplitude of which decreases gradually. If each cycle were separated into four seasons, and the peak season of economic growth would be "summer," the lowest season would be "winter," the ascending period would be "spring," and the descending period would be "autumn." Each season lasts for a period of about 2½ years.

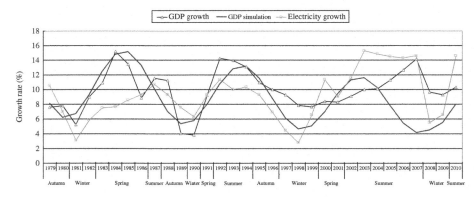

Figure 1.5 Four seasons of the Chinese economy.
Source: CEC Compilation of Statistics on Electric Power, National Bureau of Statistics.

Economic "spring" shows all the signs of spring—thriving development, booming progress. Summer describes rapid development, even to a scorching degree. Autumn is a period of economic stabilization that takes place after rapid development; in case of new conflicts and problems, adjustments need to be made. Winter represents a relatively grim economic environment, in which many challenges must be faced. The economy must be reconditioned and consolidated and wait for recovery, when it can welcome a new cycle of growth. Each season in an economic cycle has its unique characteristics and duties so as to promote the healthy development that is restarted by the economic cycle. Therefore, an economic winter is not necessarily a bad thing, and an economic summer is not necessarily good. They are both necessary stages in the process which economic development must undergo.

According to the principle of positive correlation between electric power and economy, as well as the accuracy, reliability, immediacy, and ability of electricity data to comprehensively reflect economic activity, interpreting the economy through electric power allows one to judge the cycle and season of economic development from the growth in electricity consumption.[5] The following can then be assumed: growth in electricity demand rises from 6.6% to 10% during the spring season, surpasses 10% during the summer, falls from 10% to 6.6% during the autumn season, and falls below 6.6% in the winter.

It can be seen from the electricity growth curve in Figure 1.5 that, since Reform and Opening Up, the Chinese economy has undergone four periods of downturn (1981, 1990, 1998, and 2008), forming four economic cycles.

1.3.1 Economic Cycle 1 (1981–1989)

The winter season lasted from 1981 to 1982. Although economic growth was 5.2% and 9.0%, respectively, growth in electricity consumption was only 3.1% and 5.8%. Spring was from 1983 to 1986, and during those 4 years, economic growth was 10.9%, 15.1%, 13.5%, and 8.8%, respectively. Electric power growth rose from 7.5% to 9.3%, lending unmistakable spring-like prosperity to the period. Summer was the year 1987—economic growth was 11.6%, and electric power growth was 10.7%, both exceeding 10%. Autumn came from 1988 to 1989, when economic growth dropped drastically from 11.2% to 4%, and growth in electricity fell from 9.3% to 7.5%.

The distinguishing feature of this economic cycle was that its spring season lasted 4 years, leaving only 1 year for summer. This can be described as "long spring, short summer." Figure 1.6 shows the changes in the structure of the economy. The proportion of primary industry fell from 31.9% to 25.1%. Secondary industry was the driving force of the economy at about 45%, but they still experienced a slight drop in percentage. The proportion of tertiary industry, on the other hand, rose from 22% to 32.1%, surpassing primary industry in 1985. During this cycle, electricity growth largely remained less than economic growth, and the electricity elasticity was less than 1. Figure 1.7 shows electricity intensity was in a downward state, falling 17.6% from 1633 kWh per 10,000 Yuan in 1981 to 1345 kWh per 10,000 Yuan in 1988. It did rise again to 1390 kWh per 10,000 Yuan, however, in 1989. Economic development had an obvious slant toward light industry.

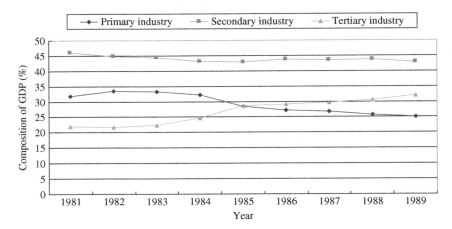

Figure 1.6 Proportion of three industries in China during Economic Cycle 1.
Source: National Bureau of Statistics.

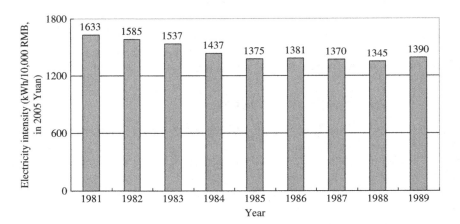

Figure 1.7 Production (per 10,000 Yuan) electricity intensity during China's Economic Cycle 1.
Source: CEC Compilation of Statistics on Electric Power, National Bureau of Statistics.

1.3.2 Economic Cycle 2 (1990–1996)

The winter season was in 1990, when economic growth was only 3.8%, and electric power growth was 6.3%. This was a short winter, but electricity growth was obviously higher than economic growth. The electricity elasticity was greater than 1, and electricity consumption rose slightly. Spring season was the year 1991 with 9.2% economic growth and 9.3% electricity growth. It was a short spring, during which the electricity elasticity was slightly higher than 1. Summer lasted from 1992 to 1994. Economic growth remained in the range of 13.1–14.3%, and electricity growth remained in the range of 10.0–11.3%. The electricity elasticity fell back below 1, and electricity consumption decreased. Autumn season was from 1995 to 1996. Economic growth was approximately 10.0%, but electricity growth fell from 9.3% to 6.9%, and the electricity elasticity continued to drop.

The main characteristic of this cycle was its short length, lasting a total of only 7 years. Looking at the structure of industry (Figure 1.8), the proportion of primary industry continued to decline to 19.7%. The proportion of secondary industry rose to 47.5%, and tertiary industry remained fairly stable in proportion with a slight increase to 32.8%. Figure 1.9 shows that the proportion of primary industry electricity consumption remained at about 5.0%. Secondary industry is a larger consumer

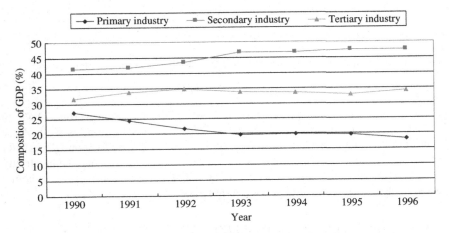

Figure 1.8 Proportion of three industries in China during Economic Cycle 2.
Source: National Bureau of Statistics.

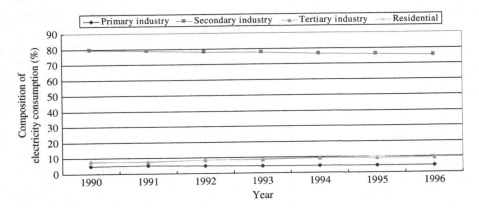

Figure 1.9 Proportion of three industries and residential electricity consumption in China during Economic Cycle 2.
Source: CEC Compilation of Statistics on Electric Power.

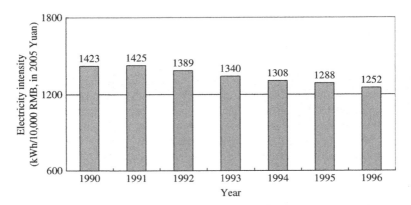

Figure 1.10 Production (per 10,000 Yuan) electricity intensity in China during Economic Cycle 2.
Source: CEC Compilation of Statistics on Electric Power, National Bureau of Statistics.

of electric power, and their proportion of consumption fell slightly from 79.4% to 75.2%. The proportion of tertiary industry electricity consumption, as well of the proportion of residential electricity consumption, gradually rose to about 10%. After a brief rise, electricity growth was for the large part in a downturn, and the electricity elasticity was less than 1. Electricity intensity fell steadily (Figure 1.10). Economic development still showed a slant toward light industry.

1.3.3   *Economic Cycle 3 (1997–2007)*

Winter lasted from 1997 to 1999. Before the Asian financial crisis began in 1997, electricity growth rate continued to drop to 4.2% which followed with the trends in the second economic cycle. In 1998, the Asian financial crisis directly impacted China's economic development, and electricity growth continued to slide down to 2.8%, a classic feature of an economy in a winter season. Spring season was from 2000 to 2001 and also was a period of economic recovery in China after the financial crisis ended. Electricity growth was higher than economic growth. Electricity growth reached 11.7% in 2000 such as entering into summer, but then fell back to 8.7% in 2001. Economic growth at the time was 8.4%. The season could be considered a "warm spring." From 2002 to 2007 was a long summer. China had just joined the WTO (World Trade Organization), coinciding with a period of relatively rapid development in the global economy, American and European economies were exceptionally well. In the context of economic globalization, demographic dividend, as well as other factors, strongly came into play, and China became "the world factory." Exports played a vital role in facilitating economic development. Except in 2002, national electricity growths achieved +14.0% and a serious shortage of electricity arose. At the same time, electricity growth was apparently higher than economic growth, and the electricity elasticity was greater than 1.

During this economic cycle, beginning in 2000, China entered into the middle phase of industrialization and then entered into the second half of this phase in 2005, and the industrialization level integrative index reached 50.[6] Industrial development was leaning toward the heavy and chemical industry. There was rapid growth in industries with high energy intensity, represented by industries like ferrous metals and nonferrous metals. The proportion of these industries in the economy rose significantly. The production of major products like cement, crude steel, and fertilizers ranked among the highest in the world. China's admission into the WTO also boosted the development of international trade. The driven force from exports grew larger and larger. From 2002 to 2007, Chinese export growth was 22.3%, 34.6%, 35.4%, 28.4%, 27.2%, and 25.7%, respectively, and import growth was 21.2%, 39.9%, 36.0%, 17.6%, 20.0%, and 20.8%, respectively. The growth of China's trade surplus surpassed 20.0% every year. Imports became an impetus for foreign economic development, and exports provided high quality yet inexpensive goods to many countries. International trade injected new life into the global economy. Nevertheless, it can also be seen that most of the products imported by China were high tech, high value added, low energy density goods, while a similar percentage of exported goods were low value added, high energy density goods. The energy density contained in the export and import products cannot be measured by only import and export trade volumes. According to estimation, deducting the energy density of imported products and the imported energy products, China's net export of energy was: 77 million toe in 1999, 59 million toe in 2000, 69 million toe in 2001, 85 million toe in 2002, 114 million toe in 2003, 151 million toe in 2004, and 222 million toe in 2005. In 2005, net export of electricity in trading was approximately 490 billion kWh.

From the structure of the economy during this economic cycle (Figure 1.11), the proportion of primary industry continued to fall, reaching 11.1%. The proportion of secondary industry, however, continually rose after 2002, reaching 48.5%.

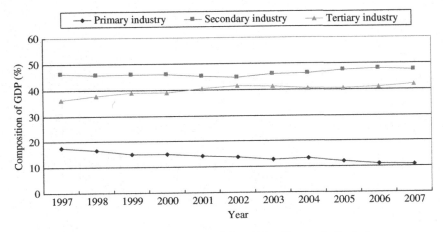

Figure 1.11 Proportion of three industries in China during Economic Cycle 3.
Source: National Bureau of Statistics.

The proportion of tertiary industry continued to rise before 2001 and remained relatively stable at about 40.0% after that. Looking at the breakdown of electricity consumption (Figure 1.12), the proportion of primary industry continued to drop to 2.65%. The proportion of secondary industry began to rise again after 2000, reaching 76.49%. After the proportion of tertiary industry rose from 9.98% to 11.21%, it fell back down to 9.78%. Residential electricity consumption rose from 11.35% to 11.45% and then fell to 11.08%. Changes in the structure of the economy caused electricity growth to consistently be higher than economic growth. Electricity intensity rose from 1129 kWh in 1999 to 1368 kWh in 2007 (Figure 1.13). Economic development showed classic signs of a slant toward heavy industries.

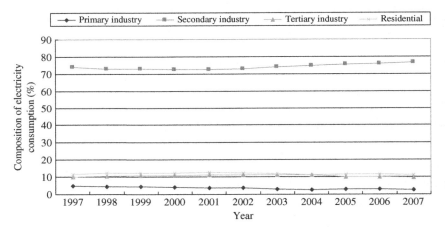

Figure 1.12 Proportion of three industries and residential electricity consumption in China during Economic Cycle 3.
Source: CEC Compilation.

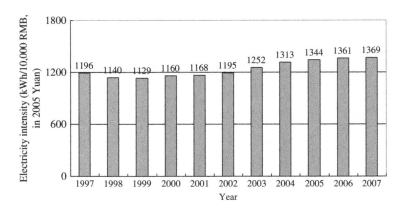

Figure 1.13 Production (per 10,000 Yuan) electricity intensity in China during Economic Cycle 3.
Source: CEC Compilation of Statistics on Electric Power, National Bureau of Statistics.

The main feature of this economic cycle was its length, which totaled 11 years, and its long summer followed by no autumn. The summer season lasted for 6 years, occupying all of autumn's time. Obviously, after 6 years of summer, the Chinese economy is worn out, and a few bubbles have appeared. The economy should have undergone an autumn season to slowly release a few of these bubbles and then a winter to provide recovery and reconditioning. However, this process was essentially omitted. Without an autumn, the economy lacked a period for recovery, and growth in electricity consumption experienced volatility. This kind of volatility is detrimental to healthy economic development and has left potential risks for healthy development in the future.

1.3.4 Economic Cycle 4 (2008–present)

Winter season was from 2008 to 2009. The global financial crisis drew the Chinese economy from a summer season straight into winter. National electricity consumption experienced negative growth from October 2008 to May 2009, and the economy sat in a state of "severe winter." In order to stimulate the economy, the Chinese government invested heavily to accelerate the recovery of economy. Electricity growth during those 2 years was less than economic growth. It was the first year in the twenty-first century that the electricity elasticity was less than 1. Economic growth was 10.3% in 2010, and electricity growth was 14.6%. If electricity growth is in the range of 6.6–10.0% for 2010–2011, then the economic cycle should enter into the spring season. Effort can be put into resolving problems and extruding bubbles in economic development in order to prepare for the 2012–2013 summer season. However, after the rapid growth in electric power in 2010, electricity consumption in 2011 was predicted to remain above 10.0%. This shows that the rapid growth in electric power in 2010 was not just an incidental phenomenon, and this was not a warm spring. Rather, the economy has once again skipped over the spring season and entered directly into summer, which will break the balance of electric power supply and demand. The approval and construction of power plants requires about 4 years to complete, which is half the length of an economic cycle. When economic growth is at its peak during the summer season, shortages in electricity occur, causing the construction of plants to be accelerated. By the time these plants go online and begin generating electricity power, the economy has already entered into the winter season, and growth in the demand for electricity has slowed, causing an excess in electricity supply capabilities and a subsequent slowdown of plant construction, which begins the entire process again. This kind of reoccurring cycle will become a spiral causing construction of electric power to be forever behind by half a cycle.

There was no spring season in this economic cycle, and there was no autumn season in the third economic cycle. These sharp swings in fluctuating economic growth will have a dramatic effect on the healthy development of the economy. During the 2009 winter season, heavy investment further inflated the real estate bubble before it could be resolved, resulting in the strange phenomenon during an economic "severe winter" in which the price of real estate not only did not fall but also rose sharply. If the real estate bubble is not controlled, it will cause a hard landing or stagflation in economic development. Bubbles can inflate an economy, but they can also destroy one. This will greatly harm the healthy development of the Chinese economy and will certainly affect future economic cycles and their seasons.

It should be seen that China's economic development still faces many problems, especially the indications that the real estate bubble has still not lessened. The central government has made great efforts to control real estate but to little effect—the price of real estate remains persistently high. Local governments have recently acted upon request from the central government, publishing directives one after another on controlling the price of real estate. Many of these in directives tie the goal of controlling real estate prices to economic growth and the disposable income of households. For example, if an area's economic growth is 10%, or the growth of its households' disposable income is 10%, then the growth of the region's real estate prices can also be 10%. This puts households in a difficult position—if they can gain benefits from economic development then the growth of households' income should be faster than economic growth. However, the faster economic growth occurs, the faster real estate prices are likely to rise, causing households to again sustain losses due to the soaring prices of real estate. The question is then whether households should welcome economic growth. The problem seems to be an economic paradox.

This phenomenon, in which the third economic cycle was without an autumn season and the fourth without a spring, is a characteristic of China's transformation process from a planned economy to a market economy. It will become an impediment to the transformation in economic development approach, and it indicates that there are still large dangers and risks in China's economic development before the industrialization process is complete (2020).

In past economic and electricity development, China's electric power elasticity has for the most part been greater than its energy elasticity (Figure 1.14), meaning that growth in electricity has been higher than growth in energy. Before 2000, economic growth for most years was greater than electricity growth (see Figure 1.5),

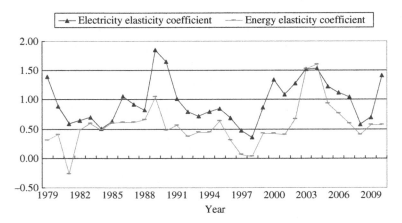

Figure 1.14 China's electricity elasticity and energy elasticity.
Source: CEC Compilation of Statistics on Electric Power, National Bureau of Statistics.

the electricity elasticity was less than 1, and electricity intensity was falling. After 2000, economic growth has largely been less than electricity growth, the electricity elasticity has been greater than 1, and electricity intensity has been rising. The year 2000 can therefore be seen as a turning point from falling to rising for electricity consumption. When then will the next turning point occur (when electricity consumption goes from rising to falling)?

1.4 Level of Electricity Consumption During Industrialization Stage

The process of industrialization is a crucial stage of economic development, economist H. Chenery evaluated economic development through examining changes within specific industrial sectors over the long term, defining modern economic growth as a comprehensive shift in economic structure, based on per capita income levels, economic growth, and structural transformation. This process of economic transformation is divided into different periods and stages as listed in Table 1.1.

The first stage is the primary production stage. The primary production is dominating the economic activities in this stage and the weight of industry in GDP is relatively small; the second stage is the industrialization stage. The stage where per capita income is greater than 728 USD (in constant 1982 $), the economic structure is shifting from primary production to manufacturing in the beginning phase of industrialization. When the contribution to the economic growth from the total factor productivity is greater than the contribution from capital growth, the nation enters the final phase of industrialization stage, and per capita income is within the range of 2912–5460 USD. The third phase is the developed economy stage, which is post-industrialization. At this point, the rate of labor productivity rises throughout the society.

As the magnitude of value changes over time, and continual changes in international flexible exchange rates and purchasing power parity (PPP), the division of economic development stages is somewhat ambiguous. According to the feature of

Table 1.1 Standard Division of Economy Development Stages (Chenery Model)

Per Capita Income (USD, in constant 1982 $)	Development Stage	
260–364	Primary production stage	
364–728		
728–1456	Industrialization stage	Early phase
1456–2912		Middle phase
2912–5460		Final phase
5460–8736	Developed economy stage	
8736–13,104		

Source: H. Chenery, Comparative study of industry and economic growth, p. 71.

positive correlation between the demand of electricity and the economy mentioned in Section 1.2, we can determine the economy stage of one nation based on their electricity consumption per capita. When a country's (region) electricity consumption is relatively large, its GDP is also large. Similarly, the larger per capita consumption of electricity, the larger of GDP per capita will be. According to Chenery's theory, 10,000 USD (in constant 2000 $, hereafter) GDP per capita is the indicator to indicate that country has completed its industrialization stage. If so, what will be the difference of electricity consumption characteristics in each stage of that country (region)?

Figure 1.15 demonstrates the relationship between electricity consumption per capita and GDP per capita in 231 countries (regions) throughout the world in 2005. In nations (areas) where GDP per capita is over 10,000 USD, electricity consumption per capita is most likely greater than 4500 kWh. Of course, we also found that some countries electricity consumption is greater than 4500 kWh, but their per capita is below 10,000 USD, anyway these individual cases still need to be analyzed specifically.

Based on the Chenery's theory, we are able to identify the different phases of industrialization by referring to the electricity consumption per capita, which in the early phase the consumption is around 300–1000 kWh per capita, 1000–2400 kWh in middle phase, and 2400–4500 kWh in the final phase; at the time of completing industrialization, electricity consumption reaches 4500–5000 kWh per capita, and the average residential electricity consumption per capita is around 900 kWh[7] (see Figure 1.2). There are two indicators to show that the nation has entered into post-industrialization stage: When a country's (region) electricity consumption per capita is greater than 5000 kWh; and residential electricity consumption per capita is greater than 900 kWh. Furthermore, the proportion of electricity consumption may become a supplementary determinant. The proportion

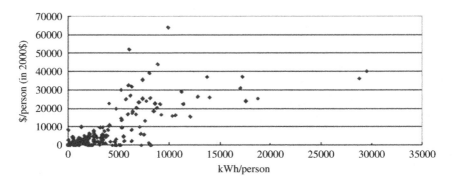

Figure 1.15 Relationship between electricity consumption per capita and GDP per capita in 231 countries in 2005.
Source: Energy Information Administration International Energy Annual 2006, http://www.eia.doe.gov/emeu/international/energyconsumption.html.

of industrial electricity consumption and increasing rate is comparatively low in the early phase of industrialization, but during the middle and the final phases the proportion and increasing rate will be rapidly increased and finally reach its peak in the post-industrialization stage growth, thereafter, gradually decreases. During the 20–40 years after post-industrialization stage, the proportion of industrial electricity consumption will decrease to 30–40%. The tertiary industry and residential electricity consumption is relatively high in the early phase of industrialization (at this phase the industrial electricity consumption weighs low) and then starts to drop; while passing to the middle and final stage of industrialization, the industrial electricity consumption will increase dramatically even entering the post-industrialization stage, the proportion will continue to increase and finally reach to approximately 30% (Table 1.2).

Fifty-six countries (regions) from 231 countries (regions) in the world are chosen to be listed in Table 1.3, where these countries GDP per capita is more than 10,000 USD or electricity consumption per capita is greater than 4500 kWh. Table 1.3 demonstrates the GDP per capita in 2005 (in constant 2000 $) and electricity consumption per capita of these 56 countries (regions). There are two parameters in Table 1.3, which "y" indicates that the GDP per capita and electricity consumption per capita both meets above condition, and "n" indicates that only one of them is satisfied. From the table you can see that there are 42 "y" which accounts 75% among the total 56 regions. In which you can also see that the GDP per capita in Trinidad and Tobago is 9673 USD and 9670 USD, 9971 USD for Malta and Slovenia, respectively. If we loosen our condition and count those three counties (regions) in our calculation, we can get a number of "80.4%" out of total 56 countries (regions).

From the table, you can see only Antigua and Barbuda, Cyprus, Portugal and Macau have GDP per capita greater than 10,000 USD but electricity consumption per capita less than 4500 kWh. Macao is just a small city, where tourism and commerce are the principal economic activities, therefore has a comparatively lower level of electricity consumption, and relatively higher GDP. It can also be seen from Table 1.3 that if the countries' (regions) GDP per capita is greater than 10,000 USD then the electricity consumption per capita is likely greater than 4500 kWh.

Table 1.2 Characteristics of Electricity Consumption in Industrialization Stage (kWh)

Economy Development Stage		Electricity Consumption Per Capita
Industrialization stage	Early	300–1000
	Middle	1000–2400
	Final	2400–4500
Completed industrialization stage		4500–5000
		Per capita residential electricity consumption: 900

Table 1.3 GDP and Electricity Consumption Per Capita of 56 Countries/Regions in 2005

Country (Region)	Electricity Consumption Per Capita (kWh)	GDP Per Capita (USD)	Markers	Country (Region)	Electricity Consumption Per Capita (kWh)	GDP Per Capita (USD)	Markers
North America	11,267	28,832.8	y	The Netherlands	5749.9	25,015	y
Bermuda	9899	63,871.9	y	Norway	29,475	40,058	y
Canada	18,760	25,344.7	y	Portugal	4117.9	11,129	n
USA	13,721	37,174.2	y	Slovakia	5492.2	4756.9	n
Antigua and Barbuda	1316.4	10,131.4	n	Slovenia	7092.7	9971.5	y
Aruba	8816.3	20,517.8	y	Spain	6746	16,872	y
Bahamas	6384.7	18,446.8	y	Sweden	17,030	30,837	y
Cayman Islands	8885.9	43,835.7	y	Switzerland	7388.1	35,540	y
The Netherlands Antilles	5329.4	13,648.6	y	United Kingdom	6156.3	27,175	y
Paraguay	7976.1	1265.3	n	Estonia	7204.6	6274.6	n
Puerto Rico	6383.1	17,592.8	y	Russia	6315.7	2404.8	n
Trinidad & Tobago	5365.1	9673.57	y	Bahrain	12 096	15,540	y
Virgin Islands	9088.3	22,527.8	y	Israel	6771.4	20,442	y
Europe	5895.1	16,936.4	y	Kuwait	17,594	24,108	y
Austria	7387	25,440.5	y	Qatar	17,224	37,158	y
Belgium	7754.7	24,211	y	Saudi Arabia	6270	8606.9	n
Bulgaria	5531.2	2175.89	n	United Arab Emirates	13,962	26,107	y

Country			
Cyprus	4070.3	10,802.8	n
Czech Republic	7535.2	5775.43	n
Denmark	6291.2	31,801.8	y
Finland	12,847	26,409	y
France	8617.3	22,851.3	y
Germany	7003.6	23,530.1	y
Greece	5237.9	14,751.9	y
Iceland	28 756	36,020.8	y
Ireland	5963.4	32,628.2	y
Italy	4791.2	19,702.7	y
Luxembourg	6050.3	51,933.7	y
Malta	5284.4	9670.22	y

Country			
Australia	11,423	22,469	y
Brunei	8496	18,421	y
Guam	10,634	16,431	y
Hong Kong, China	5238.8	30,014	y
Japan	8050	39,147	y
Korea	7619	13,321	y
Macao, China	4106.3	22,846	n
New Zealand	10318	15,956	y
Singapore	8116.2	25,765	y
Taiwan, China	9199.9	16,572	y
Global	2682.8	5671.6	

Source: Energy Information Administration International Energy Annual 2006, http://www.eia.doe.gov/emeu/international/energyconsumption.html.

1.5   Scenario Analysis of China's Economy and Electricity Demand in 2030

In order to facilitate a healthy and continuous economic growth, China will impose more macro policy and measures to strengthen the efforts to economic development transforming, accelerate the reforming of economic mechanism, actively adopt advanced technologies, enhance the efficiency of energy consumption, strengthen the environment protection in the middle and long term, and facilitate the continuous development of economy and society. Based on these factors, we have developed three possible scenarios for the increase of electricity demand and the economic development for the year 2010–2030. For the parameter values of each scenario refer to Appendix 1: Parameters for three scenarios.

1. **Scenario 1**

   In this scenario, the economic development tends to be more stabilized, labor force continuous to grow; human capital maintains accumulation, technological advance tends to be rising, and structural reforms get deeper and more thoroughly. All these changes will facilitate a more reasonable and effective allocation of economic factors among different government departments and will maintain the average annual increasing rate of TFP (total factor productivity) to be around 2% in next 20 years. Urbanization and industrialization will continue to progress. By 2020, the urbanization rate will increase to 57.6% and 64.6% by 2030. Prior to 2020, the process of industrialization shall be completed. Taking into account changes in the international economic environment and China's comparative trade advantage, trade surplus will still exist for a long period but will tend to be eliminated gradually. By 2030, the trade will be balanced and various taxes and the proportion of transfer payment will maintain the current level.

2. **Scenario 2**

   In this scenario, all kinds of mechanism reforming will be rapidly promoted and the resource allocation of market will be strengthened; the structure adjustment will be accelerated and the economic development transforming will be progressed. The government will gradually eliminate the obstacles for transferring of labor forces, the urbanization will be accelerated. Between 2010 and 2030, the annual increasing rate of urbanization will be 0.1–0.2 percentage points higher than the rate in Scenario 1. By 2020, the urbanization rate will increase to approximately 59.8% and 67.8% by 2030. The process of industrialization shall be completed by 2020. The government will increase the investment in education, health care, R&D, and social welfare; will adjust the structure of public spending and increase the ratio of government spending in education, health care, R&D and social welfare; will gradually increase the investment return of state-owned enterprises to 30–40% during 2010–2030; will increase the transfer payment from government to the poverty region and groups which will be 10–15% higher than the figure in Scenario 1; improve and complete the innovation of regulation for service industry; and gradually decrease the tax for service industry by 10%. Compare to Scenario 1, the TFP for service industry is 0.9 percentage points higher in Scenario 2; the proportion of investment in the fixed assets will vary: the investment ratio of secondary industry will be further dropped and the ratio of tertiary industry will be gradually raised. Regarding to the structure of labor force, the ratio of labor force for primary industry will be dramatically dropped and flowed to secondary and tertiary industry.

3. Scenario 3

In this scenario, we focused on the most likely possibilities, which will negatively affect the development of economy and society. In Scenario 3, the development of urbanization slows down and is 0.1–0.15 percentage points lower than the rates in Scenario 1 annually during 2010–2030, the rates will be increased to around 56% by 2020 and reach 62% by 2030. The process of industrialization shall be completed by 2020. The global economy will be undergoing a slow recovery, the trade protectionism will become more significant, and export growth will be slow. Compare to Scenario 1, the demand of export will not be recovered in a short period but will be recovered until "Twelfth Five-Years." And the export growth during 2015–2030 will be slower than in Scenario 1 due to the intensified trade protectionism, the global economic development will be slower than in Scenario 1 which will only be 2.7%. The technology innovation and efficiency improvements will also be slow down, the TFP rate will be 0.4 percentage points lower than the rate in Scenario 1. The adjustments of proportion of investment in fixed assets will not be obvious; the growth of employment rate will be slower than in Scenario 1 which will be annually 0.73% and the allocation of labor force among different industries will not change much.

Based on the simulation of above three scenarios, we are able to depict China's economic development and the growth in energy and electricity demand between 2010 and 2030 (see Table 1.4, Figures 1.16 and 1.21).[8]

Scenario 1: In 2020, China's GDP will reach 67.2 trillion Yuan, the electricity demand approximating 7.62 trillion kWh, with average GDP growth of 7.9% from 2010 to 2020, the average growth of electricity demand is about 6.2%, electricity elasticity is 0.780. The proportion of three industries accounts 6.8%, 44.5%, and 48.7%, respectively. The proportion of three industries and residential electricity consumption accounts 1.9%, 67.9%, 12.7%, and 17.5%, respectively. GDP per capita reaches 46,700 Yuan and electricity demand per capita reaches 5293 kWh. In 2030, the national GDP reaches approximately 113.7 trillion Yuan, with electricity demand of about 9.8 trillion kWh, average GDP growth is 5.4% from 2020 to 2030, and average growth of electricity demand is 2.6%. Electricity elasticity is 0.473. The economic structure of three industries is 5.8%, 39.7% and 54.5%, respectively. The proportion of three industries and residential electricity consumption accounts 1.8%, 61.3%, 15.3%, and 21.7%, respectively. GDP per capita is 77,400 Yuan and electricity demand per capita is 6672 kWh.

Scenario 2: In 2020, China's GDP will reach 68.4 trillion Yuan, the electricity demand approximating 7.73 trillion kWh, with average GDP growth of 8.1% from 2010 to 2020, the average growth of electricity demand is about 6.3%, electricity elasticity is 0.780. The proportion of three industries accounts 6.9%, 42.7%,

Table 1.4 National GDP, Energy and Electricity Demand in Three Scenarios

	Scenario	Year 2010	Year 2020	Year 2030
Scenario 1	GDP (in 2005 Yuan, trillion Yuan)	31.4	67.2	113.7
	Average GDP growth over 10 years (%)	10.46	7.90	5.40
	Primary energy demand (100 million tce)	32.50	50.71	58.65
	Average growth of primary energy demand over 10 years (%)	8.37	4.55	1.47
	Electricity demand (trillion kWh)	4.192	7.622	9.808
	Growth of electricity over 10 years (%)	12.026	6.160	2.554
	Energy intensity (tce/10,000 Yuan)	1.035	0.755	0.516
	Energy elasticity	0.799	0.576	0.271
	Electricity intensity (kWh/10,000 Yuan)	1335	1135	863
	Electricity elasticity	1.149	0.780	0.473
Scenario 2	GDP (in 2005 Yuan, trillion Yuan)	31.4	68.4	118.0
	Average GDP growth over 10 years (%)	10.46	8.10	5.60
	Primary energy demand (100 million tce)	32.50	49.86	56.93
	Average growth of primary energy demand over 10 years (%)	8.37	4.37	1.33
	Electricity demand (trillion kWh)	4.192	7.734	9.845
	Growth of electricity over 10 years (%)	12.026	6.315	2.443
	Energy intensity (tce/10,000 Yuan)	1.035	0.729	0.482
	Energy elasticity	0.799	0.540	0.238
	Electricity intensity (kWh/10,000 Yuan)	1335	1130	834
	Electricity elasticity	1.149	0.780	0.436
Scenario 3	GDP (in 2005 Yuan, trillion Yuan)	31.4	64.7	104.4
	Average GDP growth over 10 years (%)	10.46	7.50	4.90
	Primary energy demand (100 million tce)	32.50	48.37	55.75
	Average growth of primary energy demand over 10 years (%)	8.37	4.06	1.43
	Electricity demand (trillion kWh)	4.192	7.407	9.484
	Growth of electricity over 10 years (%)	12.026	5.856	2.503
	Energy intensity (tce/10,000 Yuan)	1.035	0.747	0.534
	Energy elasticity	0.799	0.541	0.292
	Electricity intensity (kWh/10,000 Yuan)	1335	1144	908
	Electricity elasticity	1.149	0.781	0.511

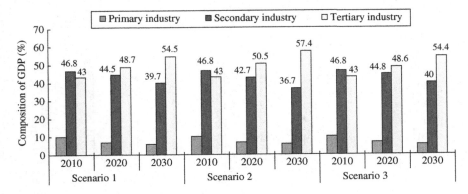

Figure 1.16 Proportion of three industries in GDP in the three scenarios.

and 50.4%, respectively. The proportion of three industries and residential electricity demand accounts 2.0%, 66.1%, 14.6%, and 17.2%, respectively. GDP per capita reaches 47,500 Yuan and electricity demand per capita reaches 5371 kWh. In 2030, the national GDP reaches approximately 118 trillion Yuan, with electricity demand of about 9.84 trillion kWh, average GDP growth is 5.6% from 2020 to 2030, and average growth of electricity demand is 2.4%. Electricity elasticity is 0.436. The economic structure of three industries is 5.9%, 36.7% and 57.4%, respectively. The proportion of three industries and residential electricity consumption accounts 1.9%, 59%, 17.5%, and 21.6%, respectively. GDP per capita is 80,300 Yuan and electricity demand per capita is 6697 kWh.

Scenario 3: In 2020, China's GDP will reach 64.7 trillion Yuan, the electricity demand approximating 7.41 trillion kWh, with average GDP growth of 7.5% from 2010 to 2020, the average growth of electricity demand is about 5.9%, electricity elasticity is 0.781. The proportion of three industries accounts 6.7%, 44.8%, and 48.5%, respectively. The proportion of three industries and residential electricity demand accounts 2.0%, 67.4%, 12.6%, and 18%, respectively. GDP per capita reaches 44,900 Yuan and electricity demand per capita reaches 5144 kWh. In 2030, the national GDP reaches approximately 104.4 trillion Yuan, with electricity demand of about 9.48 trillion kWh, average GDP growth is 4.9% from 2020 to 2030, and average growth of electricity demand is 2.5%. Electricity elasticity is 0.511. The economic structure of three industries is 5.6%, 40% and 54.4%, respectively. The proportion of three industries and residential electricity demand accounts 1.8%, 60.9%, 14.9%, and 22.5%, respectively. GDP per capita is 71,000 Yuan and electricity demand per capita is 6452 kWh.

Under the three simulated scenarios, from 2010 to 2030, primary industry's proportional share in China's GDP drops approximately 4.3–4.6 percentage points, secondary industry's proportional share in GDP drops 6.8–10.1 percentage points, tertiary industry's proportional share in GDP rises 11.4–14.4 percentage points. Also, in Scenario 2, by 2030, the secondary industry's proportional share in GDP is 57.4%, higher than Scenario 1 by approximately 3 percentage points (Figure 1.16).

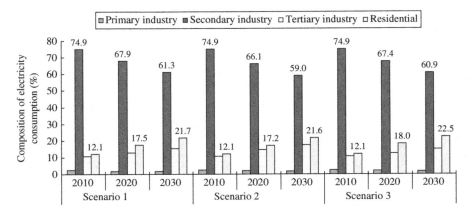

Figure 1.17 Proportion of three industries and residential electricity demand in the three scenarios.

Under the three simulation scenarios, in 2030 the proportion of three industries undergoes a significant change: the proportion of electricity demand for secondary industry will drop from 74.7% in 2010 to 61.3%, 59.3%, and 60.9% in the each scenario. The proportion of electricity demand for tertiary industry will increase from 19.7% to 14.9–17.5%; the proportion of residential electricity demand will grow dramatically from 12.2% to approximately 22% which indicates the increasing standard of living (Figure 1.17).

Scenario 2 provides a detailed simulation of economic development of eastern, central, western, and northeast regions of China and the corresponding demand for electricity (the analog of the provinces will be described in Chapters 5 and 6). The simulation shows that during 2010–2020, the fastest economic growth is in the western region with an average annual GDP growth of about 8.5%, the slowest economic growth is in the eastern region with an average annual GDP growth of 7.8%. GDP growth in central and northeast, was estimated from 8.2% to 8.4%. During 2020–2030, the eastern, central, west, and northeast regional average GDP growth was 5.1%, 5.7%, 5.8%, and 5.9%, respectively (Figure 1.18).

The eastern region's proportion of the country's economy declines gradually, whereas the central, western, and northeastern regional economies gradually increase proportionally, nevertheless the eastern region remains the economic leader. Figure 1.19 illustrates: in Scenario 2, until 2020, the east, central, west, and northeast regions constitute the following proportions of the country's GDP: 51.3%, 20.3%, 19.5%, and 8.9%, respectively; up until 2030, the eastern, central, western, and northeastern regions are estimated to represent the following proportions of the national GDP: 49.8%, 20.8%, 20.1%, and 9.3%.

The electricity demand growth was highest in the central and western regions during 2010–2020, followed by the northeast region, while the eastern region has the lowest predicted growth of electricity demand. The average electricity demand growth of the eastern, central, west, and northeast regions were 6.1%, 6.7%, 6.5%, and 6.3%, respectively. During 2020–2030, the regional electricity consumption growths will decline,

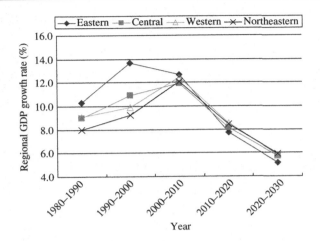

Figure 1.18 Regional GDP growth in Scenario 2.

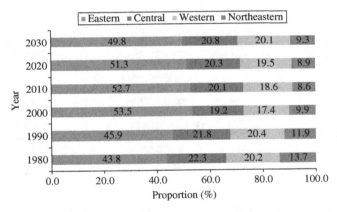

Figure 1.19 Proportion of regional GDP to national GDP in Scenario 2.

and the eastern region will be experiencing the most apparent decrease, moreover the electricity growth will be the slowest among the four regions; the western region's electricity growth will exceed that of the central region, while the northeast's electricity growth will still be lower than both the western and the central regions. During this period the estimated average growth of electricity demand in the eastern, central, west, northeast regions will be 1.8%, 2.7%, 3.5%, and 2.4%, respectively (Figure 1.20).

From the perspective of regional demand for electricity as a proportion of national demand, the proportional demand of the eastern region is significantly higher than that of the other three regions, followed by the western region, and central region, with the smallest proportional demand for electricity in the northeastern region. From a perspective on the change in proportional demand, the western region's electricity demand undergoes the most apparent increase, while the proportion of electricity demand in the eastern region tends to decline, and negligible change is noted

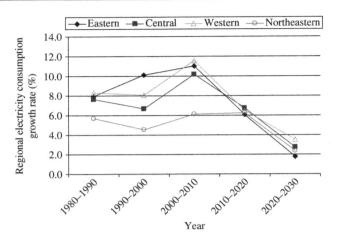

Figure 1.20 Regional electricity demand growth in Scenario 2.

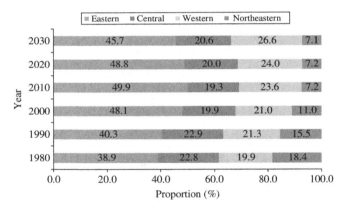

Figure 1.21 Proportion of regional electricity demand to national electricity demand in Scenario 2.

in the northeastern region's proportion of electricity demand. Up until the year 2020, the proportion of electricity demand in the country of the eastern, central, west, and northeast regions is estimated at 48.8%, 20.0%, 24.0%, and 7.2%, respectively. During the period of 2010–2020, the eastern region's proportional electricity demand will decrease by 1.1 percentage points, the proportion of electricity demand for the central and western regions will increase by 0.7 and 0.4 percentage points, respectively and the northeastern region's proportion of electricity demand will remain unchanged. By 2030, the proportion of east, central, west, northeast regions will be 45.7%, 20.6%, 26.6% and 7.1%, respectively, during the period of 2020–2030, the proportional electricity demand of the eastern region will drop by 3.1 percentage points, while the proportional demand of the central and western regions increased by 0.6 and 2.6 percentage points, respectively (Figure 1.21).

1.6 Long-Term Trends for National Economic Development and Electricity Demands

Regarding to the long-term trends for national economic development and electricity demands, we use the Scenario 1 in Section 1.5 as the basis, in ILE4, we have pre-set some parameters based on our analysis and research and simulated the demand trends of electricity based on the economic development in 2040 and 2050.

In 2040, China's total GDP will reach approximately 182 trillion Yuan (in 2005 Yuan), with a GDP per capita of about 123,700 Yuan. The average annual GDP growth will be approximately 4.8% during 2030–2040. The structure of three industries will be 1.9%, 34.9%, and 63.2%. As shown in Figure 1.22; the energy demand of primary industry is about 6.48 billion tce, with a average annual growth of 0.99% over 10 years, energy demand per capita is 4.4 tce; the total nation electricity demand reaches 12.1 trillion kWh, with an average annual growth of about 2.1% over 10 years, and electricity demand per capita is 8230 kWh. The electricity consumption proportion as shown in Figure 1.23: primary industry accounts for 1.1%, secondary industry accounts for 47.5%, tertiary industry accounts for 24.0%, and residential electricity demand accounts for 27.4%.

In 2050, China's total GDP will reach 273 trillion Yuan, with an annual growth rate at 4.16% during 2040–2050. GDP per capita is 1.87 thousand Yuan. As shown in Figure 1.24, the proportion of three industries will be 1.4%, 31.0%, and 67.6%, respectively; energy demand for primary industry is approximately 6.83 billion tce and the annual growth over 10 years is 0.54%. Per capita energy demand is 4.7 tce; total nation electricity demand is 14.3 trillion kWh and the annual growth over 10 years is 1.7%. Per capita electricity demand is 9813 kWh. The electricity proportion: primary industry is 0.9%, secondary industry is 43.8%, tertiary industry is 26.2%, and residential is 29.1% (Figure 1.25).

Table 1.5 consists of parameters such as the energy elasticity, electricity elasticity, energy intensity per 10,000 Yuan, and electricity intensity per 10,000 Yuan during

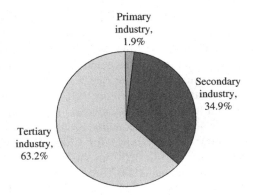

Figure 1.22 Economic structure of China in 2040.

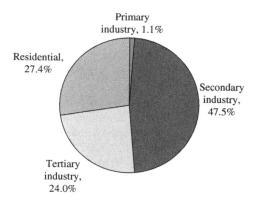

Figure 1.23 Proportion of electricity demand in 2040.

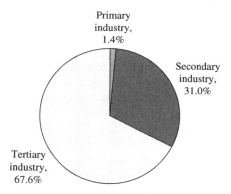

Figure 1.24 China's economic structure in the year 2050.

2030–2050. And it may be seen that China's energy elasticity and electricity elasticity are very small but competitive.

Since 1980, China already has experienced three economic cyclical stages. China's economy will continue to experience the fluctuated cyclical development. The growth tends to be more stable and the average expected rate will continue to drop. Before year 2020, the amplitude is large and the average rate is high until the completion of industrialization. The average rate will still be higher than 4% during the sixth and eighth round of cycles (Figure 1.26). Therefore, as mentioned in Section 1.2, the use of total nation electricity demand as the indicator to determine the standard of four seasons during each economic cycle needs to be adjusted.

In Figure 1.27, you may see the trends for China's economic development throughout the 70 years from 1980 to 2050: in the year 1980, China's GDP was 1.8 trillion Yuan, GDP per capita was 1800 Yuan; in the year 2000, GDP reached 11.6 trillion Yuan, GDP per capita was 9200 Yuan; in 2030, GDP will be approximately 113.7 trillion Yuan, GDP per capita will be 77,300 Yuan; in 2050, GDP

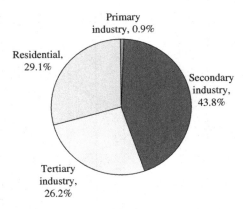

Figure 1.25 Proportion of electricity demand in 2050.

Table 1.5 China's GDP, Energy Demand and Electricity Demand During 2030–2050

Parameter	Year 2030	Year 2040	Year 2050
GDP (in 2005 Yuan, trillion Yuan)	113.7	181.8	273.2
Average GDP growth over 10 years (%)	5.40	4.80	4.16
Primary energy demand (100 million tce)	58.65	64.75	68.34
Average growth of primary energy demand over 10 years (%)	1.47	0.99	0.54
Electricity demand (trillion kWh)	9.808	12.099	14.326
Growth of electricity over 10 years (%)	2.554	2.121	1.704
Energy intensity (tce/10,000 Yuan)	0.516	0.356	0.250
Energy elasticity	0.271	0.207	0.130
Electricity intensity (kWh/10,000 Yuan)	863	666	524
Electricity elasticity	0.473	0.442	0.410
Per capita energy demand (tce)	3.99	4.40	4.68
Per capita electricity demand (kWh)	6672	8230	9813

will reach approximately 273.2 trillion Yuan, GDP per capita will be 187,000 Yuan. The average economic growth over 10 years from 1980 to 1990 was 9.28%, raised to 10.43% for 1990—2000, and 10.46% for 2000–2010. Twice of peak growth has emerged. Thereafter, it shall gradually decline to the value of 4.16% for 2040–2050. The polygonal line of economic growth showed in Figure 1.27 very much like the constellation image of the "The Big Dipper." So the question pops out, why China is able to maintain a high growth even after completion of industrialization?

As shown in Figure 1.28, you are able to see the trends for electricity demand in China from 1980 to 2050: in the year 1980, the electricity consumption was barely 3 trillion kWh; in 2000, the electricity consumption reached 1.35 trillion kWh; in

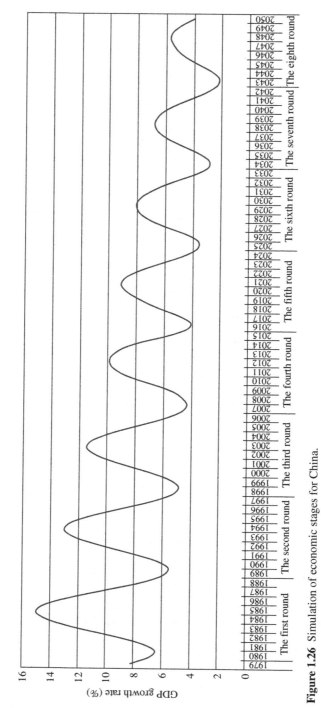

Figure 1.26 Simulation of economic stages for China.

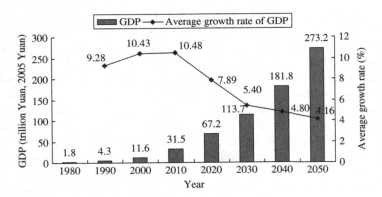

Figure 1.27 China's GDP and its growth.

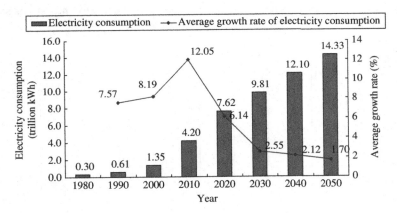

Figure 1.28 China's electricity demand and its growth.

2030, the electricity demand will be approximately 9.8 trillion kWh; in 2050, the electricity demand will be approximately 14.3 trillion kWh.

Electricity consumption in each period appears like a Chinese character "人" (Figure 1.28): the electricity consumption dramatically increased from 2000 to 2010, and thereafter experienced a sudden drop, by 2030 it will tend to be stabilized; from 2030 to 2040, the average annual growth will be 2.1%; and from 2040 to 2050, the annual growth will be 1.7%. From a perspective on the growth of electricity consumption in each stage, the annual growth from 2010 to 2020 declined substantially, in comparison with the 12% during 2000–2010, it declined by 5.9 percentage points. From 2020 to 2030, the average annual growth is lower than the previous period of 2010–2020 by 3.6 percentage points; the average annual growth of 2030–2040 is lower than the 2020–2030 period by 0.43 percentage points; and the average annual growth of the 2040–2050 is lower than the 2030–2040 period by 0.42 percentage points. A revolutionary innovation in electric vehicle technology is expected to breakthrough in 2020, and the development of electric vehicle technology will certainly influence the electricity consumption. Therefore, the growth of electricity demand will tend to be slow down after 2030. By referring to the Big Dipper graphic of China's economic growth in Figure 1.27, we can tell that China's GDP will maintain a stable development during 2030–2050 which demonstrated that the electrical vehicle technology plays a vital role in facilitation economic development. Note that this is also a unique development opportunity for China.

Electricity intensity per 10,000 Yuan is also an important indicator we need to focus on. Figure 1.29 demonstrates a downward trend for China's electricity intensity. In 2010, there was a recovery and reached 1335 kWh/10,000 Yuan. Thereafter, it will continue to decline, in 2020, the electricity intensity will be 1135 kWh/10,000 Yuan, which is 15% drop compared to 2010; in 2030, it will be 863 kWh/10,000 Yuan, which is a 24% drop compared to 2020; in 2040, it will be 666 kWh/10,000 Yuan, a 23% drop compared to 2030; and in 2050, it will be 524 kWh/10,000 Yuan, a 21% drop compared to 2040 (Figure 1.30). Due to the increasing level of electrification, the decreasing rate of electricity intensity is slower

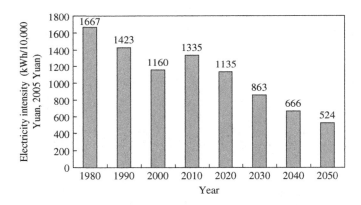

Figure 1.29 Electricity intensity from 1980 to 2050.

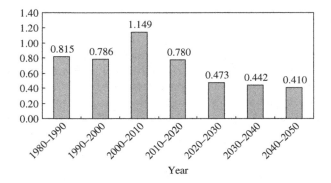

Figure 1.30 Average electricity elasticity over 10 years.

than the rate of energy intensity. Based on the statistics and research, the level of electrification and energy intensity are negative correlation. According to the characteristics of China's economic development cycle, the next inflexion of electricity intensity will be rising during 2010–2020.

1.7   Global Comparison

Through a global comparison, it is easy to see the similarity and distinctions between China and other countries on economic development tracks and trends over past 70 years and also the prediction of China's economic status by 2050. Since the statistical caliber of electricity demand differs, before we do the comparison we need to adjust China's electricity demand data based on other countries statistical caliber. Only if the statistical caliber is consistent between our nation and other countries, the conclusion shall be convincing. This section will focus on electricity demand, particularly electricity usage.

As the largest economy in the world, the GDP of the United States reached 14.25 trillion USD in 2009, GDP per capita was 46.38 thousand USD (current USD),[9] electricity consumption was 3.9724 trillion kWh, electricity consumption per capita reached 12.917 thousand kWh.[10] Figure 1.31 demonstrates the proportion of electricity consumption for industry, commerce, and resident in the United States from 1950 to 2000. The figures in first row below the horizontal axis represent the residential electricity consumption per capita, the figures in the second

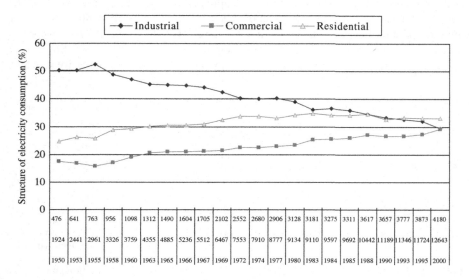

Figure 1.31 Electricity consumption per capita, residential electricity consumption per capita, and the proportion of electricity consumption in the United States.

row represent the electricity consumption per capita, and figures in the third row represent the corresponding year. In 1950, electricity consumption per capita in the United States was 1924 kWh, residential electricity consumption per capita was 476 kWh; the proportion of industrial electricity consumption accounted for 50.0%, the commercial electricity consumption was less than 17.5%, and the residential electricity consumption accounted for nearly 24.7%. In 1960, electricity consumption per capita in the United States was 3759 kWh. Although at that time the United States had completed its industrialization, due to technology applied, the level of industrial electricity consumption was not very high, so as electricity consumption per capita, but residential electricity consumption per capita was 1098 kWh. At this point, the proportion of its industrial electricity consumption was 47.0%, commercial electricity was less than 19.0%, and residential electricity consumption accounted nearly 27.2%. Along with the development of economy, per capita consumption and residential electricity consumption per capita continued to grow, while industrial electricity consumption continued to drop, by 2000 industrial electricity consumption had already fallen below 30.0% which was only 29.5%; on the contrary the proportion of commercial and residential electricity consumption was rising by 29.3% and 33.0%, respectively. At this point, its electricity consumption per capita was 12,643 kWh, while residential electricity consumption per capita was 4180 kWh.

Figure 1.32 demonstrates the average (every 10 years) economic growth, energy consumption growth, and electricity consumption growth in the United States from 1955 to 2005. The figures in first row below the horizontal axis in the following table represents the energy elasticity, the figures in second row represents the electricity elasticity, and the figures in the third row represents the corresponding year. Although the United States had already completed its industrialization, and its GDP per capita had already surpassed 10,000 USD in 1949, the United States was still able to maintain the economic growth at the level of 3–4% over 50 years, which is one aspect of its success. Moreover, the economic growth of the United States was greater than the energy consumption growth in general, the energy elasticity was less

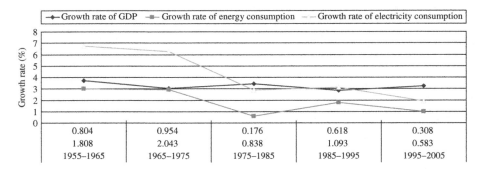

Figure 1.32 Economy, energy consumption and elasticity, electricity consumption and elasticity in the United States.
Source: http://www.eensus.gov/ and http://www.bea.doe.gov/.

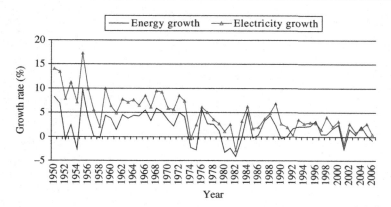

Figure 1.33 The US 1950–2006 annual growths for energy consumption and electricity consumption.
Source: http://www.census.gov/ and http://www.bea.doc.gov/.

than 1; the electricity consumption growth was consistently greater than the energy consumption growth, the electricity elasticity was greater than the energy elasticity; also the electricity elasticity was greater than 1 in general. During 1965–1975, the electricity elasticity peaked at 2.04. Due to the influence of oil crisis, the growth slowed for both energy consumption and electricity consumption in 1975–1985 and the electricity elasticity was 0.838, thereafter it recovered to 1.093 in subsequent 10 years, and it fell back to 0.583 in 1995–2005.

Figure 1.33 demonstrated the annual growths for energy consumption and electricity consumption of the United States from 1950 to 2006, from which we can easily see the electricity consumption growth is greater than the energy consumption growth, thus the electricity elasticity is greater than the energy elasticity.

Japan was a capitalist country later than the United States, by the end of the nineteenth century Japan's industrial revolution had completed, and the capitalist system had finally been established. In 1938, Japan had developed its production and manufacturing system for heavy industries such as metallurgy, chemistry, and machinery. After the World War II, Japan's economic development was divided into four periods: 1945–1956 the recovery stage: the GDP, industrial production, agricultural production, and other major economic indicators recovered to the level of Pre-War II; 1957–1973 rapid economic developing stage: the annual average of GDP growth maintain at +9%, the industrial growth was 13.6%, and Japan became the second largest economy in the world; 1974–1990 stabilized developing stage: influenced by twice oil crises, the growth slowed down and entered a more stabilized stage with average annual growth at 2.5%; and 1990s, due to the bursting of the bubble economy, Japan has experienced a 20 years period of economic depression.

Figure 1.34 demonstrates Japan's average (every 10 years) annual growth, the energy consumption growth and the electricity consumption growth rate from 1965 to 2005. The figures in the first row represent the energy elasticity, the figures in the second row represent the electricity elasticity, and the figures in the third row

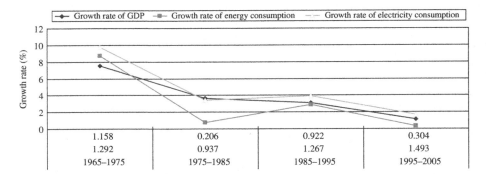

Figure 1.34 Economy, energy consumption and elasticity, electricity consumption and elasticity in Japan.
Source: The Institute of Energy Economics Japan, Handbook of energy and economic statistics in Japan, 2008.

represent the corresponding year. Japan's economy grew at 7.55% during 1965–1975, the growth of primary energy consumption and electricity consumption grew at 8.75% and 9.76%; the energy elasticity was 1.158 and the electricity elasticity was 1.292. These 10 years, Japan was in their middle stage of industrialization, was also a heavy chemical industry stage which resulted in a rapid growth of energy consumption and electricity consumption. Japan completed its industrialization during 1975–1985. At the same time, impacted by the global oil crisis the growth of energy consumption was only 0.75%; energy consumption and electricity consumption were both declining; the energy elasticity was 0.206 and electricity elasticity was 0.937. After 1985, Japan experienced an economic bubble growing and finally the bubble bursted due to the external forces in the early 1990s, Japan entered the economy depression stage. During the period of 1985–1995, Japan's energy elasticity was less than 1, and electricity elasticity was greater than 1 which indicated that even Japan experienced the economic depression in 1990s and caused its energy consumption was dropping but its electricity consumption was still rising. During the economic depression in 1995–2005, the average economic growth of Japan was 1.14%, energy consumption growth was 0.34% and electricity consumption growth was 1.7%.

Figure 1.35 demonstrates real data of annual consumption growth for energy and electricity of Japan from 1966 to 2006, from the figure you may see, in 1966, 1968, and 1969, the growth of energy consumption was greater than that of electricity consumption, with the improvements of technology and the level of electrification, Japan's growth of electricity consumption was greater than energy consumption during 1970–2006.

Figure 1.36 demonstrates Japan's electricity consumption per capita, residential electricity consumption per capita, and the structure of electricity consumption. The figures in the first row represent the residential electricity consumption per capita, the figures in the second row represent the electricity consumption per capita, and the figures in the third row represent the corresponding year; the curve shows the

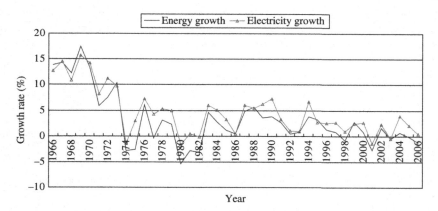

Figure 1.35 Annual growth of energy consumption and electricity consumption of Japan from 1966 to 2006.
Source: The Institute of Energy Economics Japan, Handbook of energy and economic statistics in Japan, 2008.

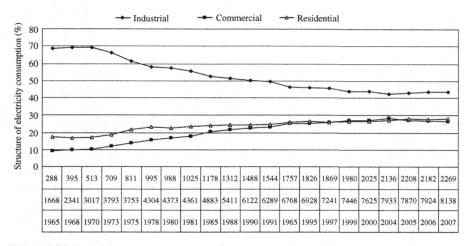

Figure 1.36 Electricity consumption per capita, residential electricity consumption per capita, and the proportion of electricity consumption in Japan.
Source: The Institute of Energy Economics Japan, Handbook of energy and economic statistics in Japan, 2008.

industrial, commercial, and residential electricity consumption proportions. In 1970, Japan was in the process of industrialization and its electricity consumption was 312.9 billion kWh, electricity consumption per capita was 3017 kWh, and residential electricity consumption per capita was 513 kWh. At this stage, the proportion of its industrial electricity consumption was 69.0%, the commercial electricity consumption accounted for 10.3%, and the residential electricity consumption accounted for

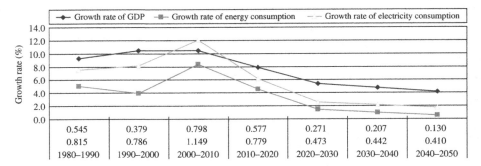

Figure 1.37 Economy, energy consumption and elasticity, electricity consumption and elasticity in China.

17.0%. Japan had completed its industrialization around 1980, and entered a post-industrialization stage thereafter. It is evident that the proportion of industrial electricity consumption continued to drop but was still above 40.0%; the proportion of commercial and residential electricity consumption did rise smoothly, both approximating 27.0%. In 2007, electricity consumption was 1.039 trillion kWh, per capita residential consumption reached 2269 kWh and electricity consumption per capita was 8138 kWh.

Figure 1.37 demonstrates the annual (every 10 years) economic growth, energy consumption growth, and electricity consumption growth of China from 1980 to 2050. The figure in the row which is below the horizontal axis represent the energy elasticity, the figures in the second row represent the electricity elasticity and the figures in the third row represent the corresponding years. It may be seen in the period of 1980–1990, the economic growth of China was 9.28%; and it maintained a growth of +10% during 1990–2010. After 30 years of rapid growth, the economy growth will be around 7.9% during 2010–2020, progressively dropping thereafter to approximately 4% in the period of 2040–2050. The electricity consumption growth curve will look like a Chinese character "人", in the past 30 years the electricity consumption rate continued to rise and in the future 40 years the growth will start to decline. For the period of 2000–2010, the growth of electricity consumption was higher than the rate of economic growth in China; the electricity elasticity was greater than 1, and the electricity elasticity was smaller than 1 during other periods, and electricity consumption was dropping. With the exception of 1990–2000, the shape of growth curve for energy consumption is similar to the one for electricity consumption. For those seven periods, the electricity consumption growth is higher than the rate of energy consumption; the electricity elasticity is greater than the energy elasticity; the economic growth is also greater than the energy consumption growth, and the energy elasticity is smaller than 1 which indicates that the energy consumption has been declining.

It may be seen from Figure 1.38: since 1979, China's electricity consumption growth was also greater than the energy consumption growth in general, but the electricity elasticity was greater than the energy elasticity.

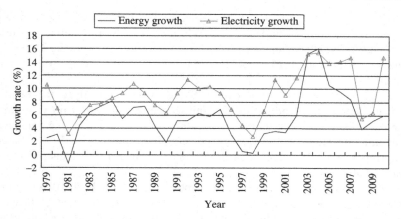

Figure 1.38 Growths of energy consumption and electricity consumption in China from 1979 to 2009.

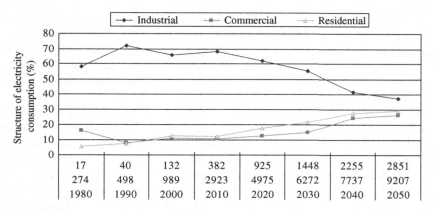

Figure 1.39 Electricity consumption per capita, residential electricity consumption per capita, and the proportion of electricity consumption in China.

Figure 1.39 demonstrates the proportion of electricity consumption for the industry, commerce, and resident in China. The figures in the first row below the horizontal axis represent the residential electricity consumption per capita, the figures in the second row represent the electricity consumption per capita, and the figures in the third row represent the corresponding years. In 1980, China was undergoing a very low electricity consumption level, electricity consumption per capita was only 274 kWh, residential electricity consumption per capita was only 17 kWh, the industrial electricity consumption accounted less than 60.0%. In 1990, the proportion of industrial electricity consumption raised to 72.3%. By 2020, the proportion of industrial electricity demand will be approximately 68%, per capita electricity demand will be 4975 kWh, per capita residential electricity demand will be 925 kWh, and thus China will be completing its industrialization.

In 2050, China's per capita electricity demand will be 9207 kWh, per capita residential electricity demand will be 2851 kWh, the proportion of industrial electricity demand will be approximately 43.8%, the proportion of commercial electricity demand will be around 26.2%, and the proportion of residential electricity demand will be 29.1%.

From a comparison with the United States, you can analyze the characteristics of economic development in the United States over nearly 60 years after it completed industrialization; and from a comparison with Japan, we can observe its economic development, electricity consumption together with its characteristics during 1965–2005 which was the period for Japan from industrialization to post-industrialization.

The level of China's electricity consumption per capita in 2010 is equivalent to the level of the United States in 1955. In 1955, the United States already had completed the process of industrialization, GDP per capita was 13.4 thousand $ (in constant 2000 $), electricity consumption per capita was 2961 kWh, electricity consumption per capita was 763 kWh. Regarding the structure of electricity consumption, its industrial electricity consumption accounted for 52.3%, commercial electricity consumption accounted for 15.9%, and residential electricity consumption accounted for 25.8%. In the period of 1955–1965, the United States experienced a sharply dropping of residential electricity consumption, and a rapid increase of electricity consumption for the commerce; its average GDP per capita rose by 23% and average electricity consumption per capita rose by approximately 65%, and the electricity elasticity was greater than 1.

The level of China's electricity consumption per capita in 2010 is equivalent to the level of the Japan in 1970. In 1970, Japan was undertaking its industrialization stage, GDP per capita was 19.174 thousand Yen, electricity consumption per capita was 3017 kWh, residential electricity consumption per capita was 513 kWh. Regarding the structure of electricity consumption, its industrial electricity consumption accounted for 69%, commercial electricity consumption accounted for 10.3%, and residential accounted for 17%. In the period of 1970–1980, its electricity consumption for commerce rose to 16.8%, and the residential electricity consumption dropped to 57.5%; its average GDP per capita rose by 36% and average electricity consumption per capita rose by approximately 45%, and the electricity elasticity was greater than 1. By 1980, Japan had completed its industrialization.

This also enlightened us, the commerce is always gaining opportunity in fast growth during the completion of industrialization process. In 2010, China's commerce only constituted a proportion of 10.73% of total electricity consumption, and there was still a lot of room for growth. From 2010 to 2020, China will be undergoing the completion of industrialization stage, the commerce will also face a lot of development opportunities. Due to technology's improvements and government's macro policies imposed on energy conservation and emissions reducing, as well as the supporting efforts will be given to managing the electricity demand side, China's electricity elasticity will be possibly smaller than the level of China's electricity consumption per capita in 2020 will be equivalent to the level of the United States in mid-1960s. In 1965, the GDP per capita of the United States was

16.5 thousand USD, electricity consumption per capita was 4885 kWh, residential electricity consumption per capita was 1490 kWh. Regarding the structure of electricity consumption, its industrial electricity consumption accounted for 45%, commercial electricity consumption accounted for 21%, and residential accounted for 30%. In the period of 1965–1975, its residential electricity consumption continued to drop, and the electricity consumption for the commerce continued to rise; its average GDP per capita rose by 21.3% and average electricity consumption per capita rose by approximately 64.5%, and the electricity elasticity was still greater than 1.

The level of China's electricity consumption per capita in 2020 will be equivalent to the level of Japan in mid-1980s. In 1985, the GDP per capita of Japan was 29.37 million Yen, electricity consumption per capita was 4883 kWh, residential electricity consumption per capita was 1178 kWh. Regarding the structure of electricity consumption, its industrial electricity consumption accounted for 52.65%, commercial electricity consumption accounted for 20.37%, and residential accounted for 24.12%. In the period of 1985–1995, its economy increased by 1.358 times, national electricity consumption increased by 1.89 times, and the growth of electricity consumption is higher than the rate of economic growth.

Based on the same level of electricity consumption and in terms of the economic development trends, the economic growth of the United States was 3.05% during 1965–1975, electricity consumption was 6.24%, and the electricity elasticity was as high as 2.04; while Japan experienced a recession after rapid growth during 1985–1995, but its electricity consumption growth was still not low, and the electricity elasticity was still greater than 1. This drew our attention, we must unswervingly strengthen the demand side management, ensure a drop of electricity elasticity during 2020–2030.

In 2030, per capita electricity demand of China will reach 6272 kWh, per capita residential electricity demand will reach 1448 kWh, which are lower than the rates of the United States in 1969 (6467 and 2102 kWh) and is equivalent to the level of Japan in 1991 (6289 and 1544 kWh). It took Japan 6 years (1985–1991) to increase its electricity consumption per capita to 1406 kWh and will take China 10 years (2020–2030) to increase per capita electricity demand to 1297 kWh.

In 2040, China's per capita electricity demand will reach 7737 kWh, and per capita residential electricity demand will reach 2255 kWh, which are lower than the rates of the United States in 1974 (7910 and 2680 kWh) and is approaching to the level of Japan in 2005 (7870 and 2208 kWh). And it will take China another 10 years (2030–2040) to reach the level of per capita electricity consumption of the United States and Japan. Note: the United States took 5 years (1969–1974) and Japan took 14 years (1991–2005).

In 2050, the electricity demand of China will be 9207 kWh, which will be almost the level of the United States in 1980 (9134 kWh). The United States is rich in land, abundant natural resources, and electricity consumption per capita is high. China will not be able to reach the level of the United States. However, in 2050, China's per capita electricity demand will surpass that of Japan in 2008 (approximately 8200 kWh). And the key reason for this is that after 2030, the electric vehicles may undergo a rapid development and as a result it will increase the electricity demand of China.

From a comparison with the United States, we can see that with the restriction of technology and electrification development, even the United States had already completed its industrialization before 1950, its electricity consumption per capita was also low. As technology continued to improve, GDP per capita also continued to grow quickly, electrification rapidly developed and its electricity consumption per capita also grew quickly. Moreover after 2020 China's economic growth will be slowed down. With the advancements of production technology and equipment, a highly electricity efficiency can be achieved and the electricity demand will drop sharply. This is why China possibly can maintain an electricity elasticity lower than 1 onward, and why it will take China 10 years time to achieve the growth of electricity consumption of the United States with 5 years.

From a comparison with Japan, we can see that with the improvements of technology and adoption of measures in the management before and after its industrialization, China will facilitate its energy saving and these will endow China advantages to maintain its energy elasticity and electricity elasticity to be the lowest. Particularly in the period of 2030–2040, China's energy elasticity will achieve 0.207, and the electricity elasticity will achieve 0.442 which are far lower than the rates of Japan's in the period of 1995–2005 (0.304 and 1.493, respectively) and the rates of the United States in the period of 1970–1980 (0.446 and 1.308). In the period of 2040–2050, China's energy elasticity will be 0.130, and the electricity elasticity will be approximately 0.410, which are far lower than the United States in the period of 1980–1990 (0.274 and 0.946). China's electricity elasticity will be outstanding in the world, and we all know that this is achievable and expectable.

Of course, it must be noted that to achieve this level, the challenges are considerable. If we failed to cope with different challenges, the electricity elasticity will not drop substantially and the electricity demand may also be higher than the above analysis in the period of 2040–2050. The objectives will become expectable but not achievable.

Is the given challenges not realistic? And have we underestimated our electricity consumption? China is a country with a large population which will reach 1.46 billion people in 2050; resources are not abundant, and the per capita resources are far below the world level. China is facing high pressure on the environment issues. Above factors are all the key obstacles to our economic development. Electricity consumption per capita will reach 6272 kWh in 2030, and 7737 kWh in 2040, which is equivalent to the consumption level of developed countries. If we do not consider the breakthrough of electric vehicles and many other electricity technologies, China's electricity demand per capita will be saturated around 8000 kWh. However, the development of electric vehicles will push China's per capita electricity demand to 9207 kWh in 2050.

Of course, the above analysis is based on specific parameters settings, different parameter settings will yield a different analytical conclusion.

Notes

[1] The GDP listed in the National Bureau of Statistics' 2010 Statistical report on the development of the National Economy and Society on February 28, 2011, was 39.79 trillion Yuan.

[2] China Electricity Council, Electric power statistics bulletin, February 2011.

[3] Zhaoguang Hu. Looking at economy through electric power-important research of macroeconomic control. Modern Electricity 2005;22(2):1–6.

[4] Zhaoguang Hu. Analysis of Chinese economic development and trends in electric power demand. Electric Power 2000;(8):6–9.

[5] Zhaoguang Hu. Considerations of China's 'twelve year plan' comprehensive strategic planning for resources. Energy of China, Experts' Forum 2009;(9):12–14.

[6] Chen, Jiagui, et al. The report on Chinese industrialization. Beijing: Social Sciences Academic Press; 2007, p. 47.

[7] Hu Zhao Guang. Primary analysis of the influence of electricity usage on my nation's economic development and electricity power demands. China's Energy 2007;29(10).

[8] National and regional economic simulation through the year 2030 carried out by the State Council Development Research Center using the DRC-CGE Model.

[9] IMF, World Economic Outlook Database, 2010.

[10] IEA, Energy Balances of OECD Countries, 2010.

2 An Introduction into the Intelligent Laboratory of the Economy–Energy–Electricity–Environment (ILE4)

2.1 Intelligent Laboratory

ILE4 is essentially a soft science laboratory which specializes in economy, energy, electricity, and environment. It uses Intelligent Engineering theories and methodologies, along with modern computing technology, to conduct research and analysis into topics that cover: economic development, energy demand, electricity demand, the relationship between electricity usage and the economy, as well as the electricity supply–demand equilibrium and warning indicators.[1] It handles these tasks at both the national and regional level.

What is a soft science laboratory? It is similar to a natural science laboratory. A soft science laboratory is made up of the following structure:

Soft science laboratory = tool base + research platform

Within the tool base, there are methodologies, models, data and related knowledge which are useful for research. This structure is detailed below:

Tool base = methodology base + model base + knowledge base

In fact, in soft science research, one can use the tools from the tool base for activities on the research platform, while still also observing the entire process from beginning to end. The process is detailed below:

Soft science laboratory = person + tool base + research platform

All experiments are conditional. That is, if the X condition is set, then the Y result occurs. This is known as "if... then..." reasoning. Prediction also uses this type of reasoning. A predictive outcome or "result" is a consequence of whatever preliminary "conditions" are set. For prediction, one can conceive of many possible "conditions." These "conditions" naturally lead to many corresponding "results." They also yield a large variety of possible scenario analyses. In such research, one needs assistance from soft science laboratories. Researchers then set up many preliminary "conditions" and use the research platform to observe the "results." ILE4 provides researchers with these types of research platforms.

In the first part of the book, characteristics of electrical power, the high positive correlation between electricity demand and economic output, and aspects of

An Exploration into China's Economic Development and Electricity Demand by the Year 2050.
DOI: http://dx.doi.org/10.1016/B978-0-12-420159-0.00002-3

intelligent engineering theory were introduced. These things constitute the theoretical foundation on which ILE4 works.

How can we express the relationship between electricity demand and economic development? People once used simple mathematical models, such as regression models, to express this relationship. More recently, along with an increasingly deeper understanding of the problems, more complex mathematical models (neural networks, system dynamics, etc.) have started to appear. Due to the complexity of economic systems, which are characterized by nonlinearity, dynamic trends, higher dimensional parameters, and uncertainty, many problems cannot be expressed simply by means of mathematical models. Generalized models are an extension of traditional mathematical models. They include regulated models, fuzzy inference models, as well as neural networks and mixed models, just to name a few. By utilizing generalized models, intelligent space, and the intelligent paths at the core of intelligent engineering theory, we can approach the above research using new methods.[2–4] Intelligent Engineering is a supplement to an expansion of systems engineering. It upgrades mathematical modeling in systems engineering to more generalized modeling. Furthermore, it extends state space to intelligent space and replaces the state space's state transition matrix requirements with solutions for intelligent paths within the intelligent space. Intelligent engineering theory is illustrated in Figure 2.1.

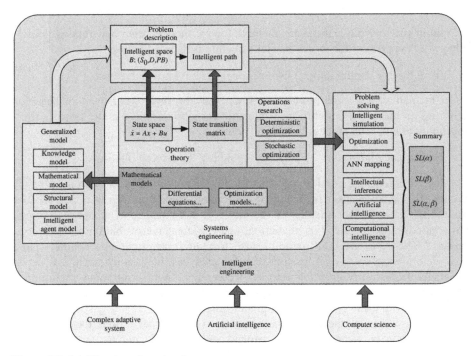

Figure 2.1 Intelligent engineering theory.

Using both intelligent engineering theory and methodology, ILE4 examined questions concerning macroeconomics, energy supply and demand, meteorology and hydrology, the connection between electrical power and the economy, the electrical power supply, electrical power demand, and the electrical power supply–demand equilibrium. Within ILE4, researchers can come to understandings about national and regional economic development and electricity supply and demand conditions. They can also analyze different characteristics of economic development and electricity supply and demand. They can research national and regional economic development and their connection with energy/electrical electricity demand, analyze future economic development, as well as electricity supply and demand. Furthermore, they can examine early warning indicators for potential problems with the national economy or electrical power operations and conduct simulation analyses on macroeconomic policy and the national economy to understand their impact on electricity supply and demand. This type of research can contribute to decision making in government policy.

The framework of ILE4 is illustrated in Figure 2.2, which starts with an analysis of macroeconomic situations, electricity supply and demand, and the relationship between electrical power and the economy. From here, predictions about future power supply and demand are made after taking into account the economic trends. Furthermore, the influence of energy supply and demand, meteorology and hydrology, and other factors are also incorporated. For such forecasts, both qualitative and quantitative methods are used. Based on this initial predictive process, power supply and demand equilibrium analyses, and early warning analyses on electrical power supply and demand, are carried out. According to the forecasts, relevant policies are revised, and simulations are analyzed to see the effects of policy changes. All of this will hopefully lead to policy optimization. ILE4's generalized models, expertise, data, and powerful computer-processing capabilities are combined. Then, the closed-loop research process which consists of "analysis forecasting–early warnings–simulation–expert discussion–analysis forecasting" is implemented.

2.2 Primary Functions

ILE4's primary functions are as follows:

1. Macroeconomic performance analysis and forecast

    Macroeconomic performance analysis and forecast can lead to an analysis of all types of macroeconomic performance indicators. This allows for the forecasting of future economic trends and provides support for forecasts on electricity supply and demand. The principle components of the analysis and forecasts are gross domestic product (GDP), economic structure, the development of key industries, investment, consumption, imports and exports, production volumes and prices of key goods, and the performances of key industries.

2. Energy supply and demand analysis and forecasting

    Energy supply and demand analysis and forecasting can be used to analyze both current and historical data for national and regional primary energy production. It can also be used

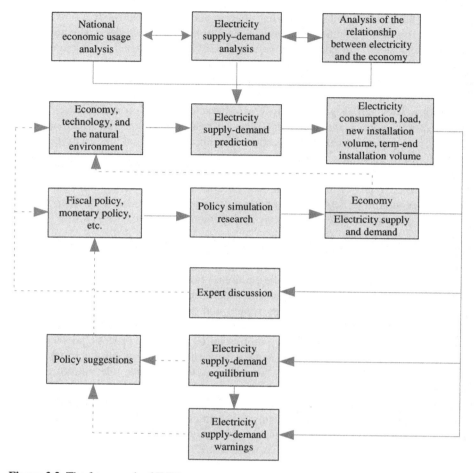

Figure 2.2 The framework of ILE4.

to analyze primary energy demand and final energy (coal, oil, natural gas, and electric) demand. Furthermore, it is capable of forecasting future energy demand.

3. Meteorological and hydrological analysis

A meteorological and hydrological analysis can be achieved for maximum temperature, minimum temperature, average temperature, average relative humidity, precipitation quantity, average wind speed, and other trend indicators. It can be carried out at both national and regional levels.

4. Analysis of the relationship between electricity and the economy

An analysis of the relationship between electricity and the economy allows for a comparison between the relationship of electricity consumption and economic growth at both national and regional levels. Furthermore, one can also calculate and analyze the index of energy elasticity, index of electricity elasticity, energy consumption by GDP, and electricity consumption by GDP. Furthermore, analysts can also apply structural decomposition models in order to conduct a decomposition analysis on changes in energy/electricity intensity.

5. Electricity supply and demand analysis and forecasting

An electricity supply and demand analysis consists of an electricity supply capacity and demand analysis. The electricity supply and demand analysis covers installation capacity, generated electrical energy, and the hourly use of electrical equipment. An electricity demand analysis covers: total national and regional electricity demand, subindustry power use, maximum power loads, and power load characteristics. An analysis of electrical power load covers: daily loads, monthly loads, annual loads, summer (winter) maximum load dates, typical daily load curves, summer (winter) workday (holiday) air conditioning loads, annual load curves, annual load duration curves, and so on.

Electrical power supply and demand forecasting include electricity supply capacity forecasting, electricity demand forecasting, and load forecasting. It can make forecasts about national and regional electricity demand, as well as industry electricity demand, electrical loads, load curve simulations, and end-of-term installation capacity.

6. Electricity supply and demand equilibrium analysis

An electricity supply and demand equilibrium analysis primarily consists of an analysis on the electricity supply–demand equilibrium within the national, regional, and provincial (region, city) grids. According to forecasts about the end-of-term installation capacity and electricity load and usage, and by taking into account maintenance and electricity exchange, calculations are then made regarding regional or provincial (region, city) power grid usage rate, hours of use by generating equipment, and major electricity shortages or surpluses.

7. Electricity supply–demand warning indicators

Electricity supply–demand early warning functionality is based on the electricity supply capacity and the electricity demand forecast. It uses early warning methods to determine future electricity supply and demand, as well as the plumbing conditions of supply–demand equilibriums. The indicators supply warnings about potential electricity surpluses or shortages. In ILE4, electricity supply–demand warnings mainly use electricity supply–demand indexing. These indexes are divided into annual and quarterly electricity supply–demand indexes. Annual supply–demand indexing provides subannual, national, and regional electricity supply–demand warnings, while quarterly electricity supply– demand indexing provides subquarterly regional electricity supply–demand warnings.

8. Policy-simulation research

Policy-simulation research allows for the analysis of various macroeconomic policy influences on economic development, electricity supply, and demand. Here, research tools include agent-based and computable general equilibrium (CGE)-based policy simulations. Agent Response Equilibrium (ARE) models can simulate fiscal policy, monetary policy, industrial policy, and other influences on the economy, as well as electricity supply–demand. CGE-based policy simulations can be used to analyze exchange rates, taxes, and other policy-related influences on the economy and electricity supply–demand.

9. Integrated discussion

Comprehensive integrated discussion, which presents a methodological application of system integration, keeps key problems in mind which the electricity supply–demand analysis and forecasts are mandated to solve. Issue quantification is achieved with the assistance of several different tools. These tools include Internet communications technology, databases, data storage technology, computer simulation analyses, multimedia technology, collaborative working techniques, as well as fuzzy decision making, combined with expertise and strong computer-processing capabilities.

10. Geographical information display

Geographical information display gives geographical information for all economic, energy, electricity, meteorological, and hydrological data. It allows for the calculation of various proportions (total proportion, partial proportion), growth rate, increment, pull rate, and contribution rate for all indexes. With geographical information databases, it is possible to conveniently manage all kinds of electrical system installations (transformer substations, power plants, overhead lines, etc.). We can also carry out statistics gathering, analysis, and data summaries.

11. Searching through macroeconomic policies

This functionality allows for the analysis of economic, energy, and electrical policies. It also has implications for the related laws and regulations, which are primarily made up of fiscal, monetary, industry, energy, environmental, trade policies, laws and regulations. By searching and browsing, one can come to a sound understanding of national policies. This provides support for a future policy-simulation analysis.

2.3 Primary Models

ILE4 has more than 30 types of electricity supply–demand forecasting models. These include linear and nonlinear regression models (exponential curves, inverted exponential curves, power function curves, logarithmic curves, hyperbolic curves, and S-curves). Using a time series, one can create moving average models, exponential smoothing models, exponential curve models, logistic models, error correction models, vector auto-regression models, and auto-regression distributed lag models that have also been established according to econometric methods. "Top-down" and "bottom-up" thinking is responsible for the generation of LEAP and other models. Input–output theory has also led to the establishment of input–output models, while Intelligent Engineering technology has led to the development of agent-based models. In addition, there are also neural network models, gray models, fuzzy inference models, per capita electricity demand models, unit demand models, coefficient of elasticity models, regional decomposition models, combined models, and so on (Figure 2.3).

In addition to the electricity supply–demand forecast model, ILE4 can design and develop specialized models in order to serve various other research objectives. For example, as far as macroeconomic performance analysis is concerned, macroeconomic monthly models, macroeconomic mid- and long-term models, production-function models, and CGE models have been developed. In terms of an analysis on the relationship between electricity and the economy, singular value decomposition models have been developed. Based on the electrical power supply, integrated resource strategic planning (IRSP) models have been developed. Furthermore, based on analyses of the electricity and energy balance, computational, regional, and provincial (region, city) grid electricity and energy balance models have been developed. Based on electricity supply–demand warnings, annual electricity supply–demand index, quarterly electricity supply–demand index, electrical power industry boom analysis, and other models have been developed. In terms of policy simulations, agent-based policy-simulation models have been developed. Finally, in order to investigate the relationship between electricity and the economy, structural decomposition models have been developed.

Below, we have introduced some macroeconomic performance analysis models, as well as some electricity supply–demand forecast models.

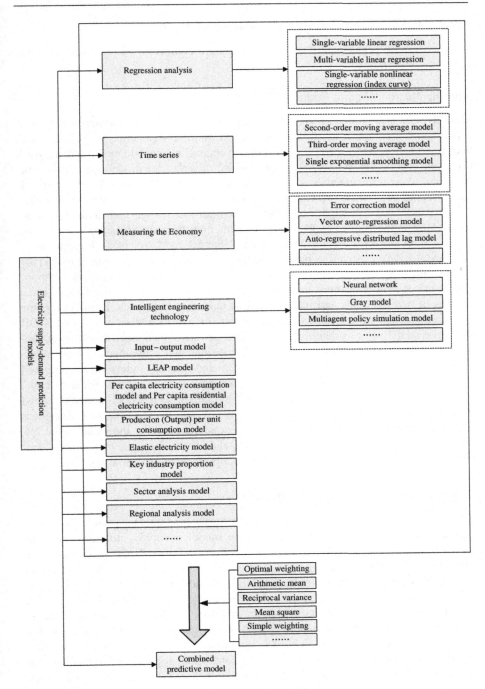

Figure 2.3 ILE4 electricity supply–demand forecast models.

2.3.1 Macroeconomic Performance Analysis Models

Macroeconomic performance analysis models are comprised of the five following types.

Monthly Macroeconomic Model

Monthly macroeconomic model is constructed according to macroeconomic and econometric theories. It usually includes eight major components, which are fiscal budget, industry, trade, currency, prices, investment, resident income, and national economic accounting. With 47 equations and 82 variables, the model is used for some mid- and short-term macroeconomic analyses. It may also be used for sensitivity and dynamic analyses of mid- and short-term influences on primary economic factors within China. The eight components are both interrelated and mutually conditional. GDP is determined by capital investment, consumer spending, and net exports. Capital is determined by resident savings, interest rates, overseas foreign direct investment (OFDI), as well as budget structure and governmental consumption. Trade is determined by the global GDP growth rate, exchange rates, and primary import/export prices. The consumer price index (CPI) is determined by the factory price indexes of raw materials and fuel, as well as manufactured goods. Resident income is determined by governmental consumption, industrial added value, and GDP. Industrial added value is determined by exchange rates, fixed asset investments, factory price indexes on manufactured goods, governmental spending, M2, and industrial enterprise sales taxes and surtaxes. Banking and currency indexes are determined by exchange rates, industrial added value, interest rates, prices, and fiscal revenue. Fiscal revenue and expenses are determined by enterprise revenue, industrial added value, imported equipment cost, nonferrous metals, and the steel industry. The theoretical framework underlying monthly macroeconomic model is shown in Figure 2.4. The left side represents exogenous variables, and all of others are endogenous variables. The arrows indicate the logical relationships between the variables.

Mid- and Long-Term Macroeconomic Model

Mid- and long-term macroeconomic model owe its fundamental principles to neoclassical economic growth theory. Long-term model is supply oriented and based on the annual data from 1990 to 2009. It is constructed using linear regression, nonlinear regression, time series analysis, and other techniques. The model can forecast mid- and long-term macroeconomic development, industry structure, primary energy production and consumption, as well as the investment and consumption rates. The model includes six components, which are industry, residents, government, energy, trade, wages, and prices. It contains a total of 49 equations and 105 variables. The exogenous variables consist mainly of taxes, population and proportion of industrial employment, interest rates, exchange rates, global GDP, foreign investment, and fiscal budgets. The endogenous variables primarily include GDP (and its actual growth rate), industry added values (and their actual growth rates), total fixed asset income, resident and household income, energy consumption and production, revenue and

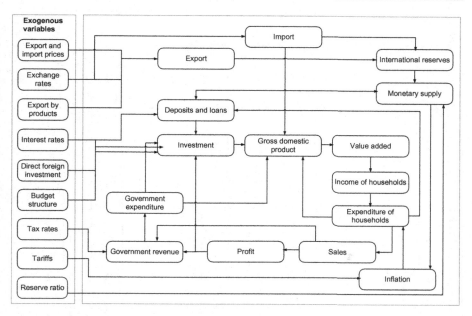

Figure 2.4 Principle of monthly macroeconomic model.

expenditures, import and export volumes, and price indexes. The theory of mid- and long-term macroeconomic modeling is shown in Figure 2.5. In Figure 2.5, the left side contains the necessary exogenous variables, while the others are some primary endogenous variables. The arrows indicate the interrelationship between the variables.

This model uses production functions to portray the production processes of the secondary and tertiary industry. Labor input is measured by year-end employment numbers across sectors. Capital investment is calculated on the basis of various industrial fixed asset investments. In this model, the total fixed asset investment value is decided by secondary industry added value, foreign investment, enterprise three-year deposit rate, and the investment price index. Resident income and consumption are decided by secondary and tertiary industry salaries, primary industry added value, and the three-year resident deposit rate. Government account is decided primarily by tertiary industry added value, value-added tax (VAT), turnover tax, and fiscal budget. Energy consumption is decided by GDP, resident consumption expenses, and the historical growth in energy consumption. Foreign trade is decided by exchange rates, domestic demand, and the import price index. Wages and price indexes are decided by average salary, GDP indexes, nonenergy price indexes, and exchange rates.

Solow Model

Long-term economic growth is mainly determined by factor productivity, capital, and labor input. Some economists use cleverly structured simple model systems

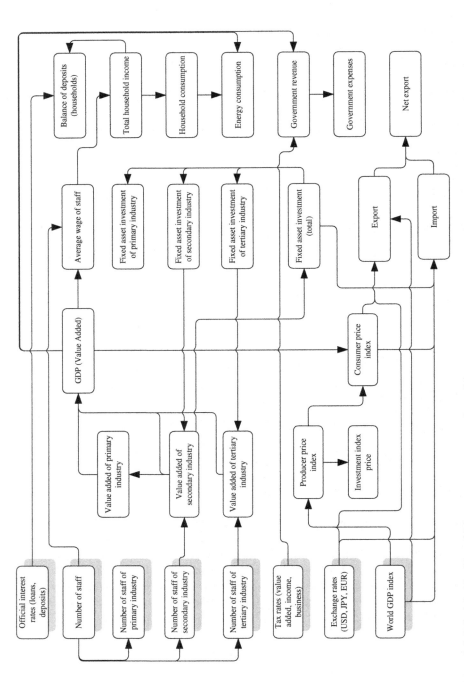

Figure 2.5 Principle of mid- and long-term macroeconomic model.

to perform quantitative analyses on the factors which influence economic growth. Furthermore, the trends in these factors are then analyzed. Therefore, by adopting a simple model that includes a few mathematical equations, they can then make rational forecasts about long-term economic growth.

ILE4 sometimes uses the economic growth model originally created by Robert Solow. It is known as Solow model.[5,6] The model is comprised of four equations.

The first equation, the Cobb–Douglas production function describes overall economic growth by the means of constant returns to scale. Assuming that GDP is decided by supply side factors, it is calculated as:

$$Y_t = AK_t^\alpha L_t^{1-\alpha}$$

Y represents the GDP, A represents the total factor productivity (generally used to represent technological progress), K represents the supply of capital, L represents the labor supply, and α and $1 - \alpha$ represent the capital and labor output elasticity, respectively.

The second equation expresses changes in capital using the perpetual inventory method. The calculation is:

$$K_t = (1 - \delta)K_{t-1} + I_t$$

In this equation, δ represents the rate of deprecation, and I is the net investment.
The third equation expresses change in the workforce, with the formula:

$$L_t = aL_{t-1}$$

In the equation, a is the labor growth rate.
The fourth equation depicts savings and is expressed as:

$$I_t = bY_{t-1}$$

Herein, b represents the investments accounted for on a proportion of GDP over the last period.

Computable General Equilibrium Model

CGE model is a model that turns the abstract GE theory to a real economy situation. The basic principle of CGE model is shown in Figure 2.6. Using a single group of mathematical equations, the model describes the equilibrium relationship between supply and demand on various markets. In this group of equations, there are exogenous variables (generally showing shocks received by the economic system) and endogenous variables (generally showing quantities and prices of goods within the economic system). Changes in the exogenous variables that influence any part of the economy can spread to the entire system. This leads to universal changes in the quantity and prices of key goods and factors. These changes can cause the entire economic system to shift from one state of equilibrium to another. The CGE model can calculate a set of figures and prices when the demand and supply reach equilibrium during the transition period.[7]

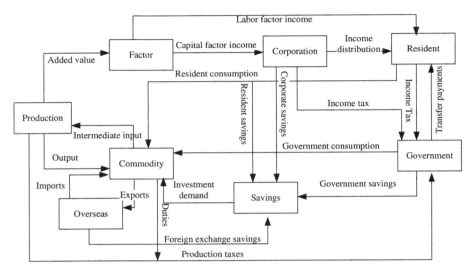

Figure 2.6 Diagram of CGE modeling.

In Figure 2.6, production activity, subjects, goods, essential factors, etc. are abstracted to eight types of accounts. These are shown in the boxes. The arrows between the boxes show the cash flow between accounts. The arrows which point into an account represent income going into that account, while the arrows that point out of an account represent expenses paid to another account. This figure describes the economic cycle; production leads to income, income leads to demand, and demand leads to production.

The accounts within the CGE work as follows: the production account shows production activity within the economic system. The factor account contains production factors, consisting of capital and the labor force. The enterprise account contains corporations, including agricultural, industrial, service, and other industries. The resident account contains residents (both urban and rural). The government account contains branches of the government, including both central and regional governments. The merchandise account portrays the production of various goods, including agricultural, industrial, and service goods. The foreign account represents countries other than China. China, on the one hand, wants to export goods, and, on the other hand, wants to import some foreign goods. The savings account contains deposits, which include the deposits of: residents, governments, corporations, and foreign entities. This account also contains the savings which have been shifted to investment.

Residents acting as owners of labor and corporations acting as owners of capital, provide the factor inputs into production. Production uses primary factor and intermediate input to produce goods. In this process, value is added which becomes profit within the factor account. In the next step, this factor profit is distributed to proprietors. Labor income goes to the residents, while capital profits go to corporations. Corporations also distribute a part of their income to residents. Governments act as agents of economic regulation. Government income comes from production taxes, sales taxes, resident and corporate income taxes, and import duties. Government income is sometimes used in transfer payments to businesses and residents. Both residents and governments use their income for the consumption of goods. Besides consumption, the remaining income is used for savings. Corporations, besides taxes and payments to residents, also use their income for savings. Goods that are not produced domestically come from foreign imports, and those that are not sold domestically can be sold internationally. A foreign account surplus, resulting from international trade, makes up one part of total deposits. Deposits are used to purchase capital goods by means of investment and become new capital factor inputs in the next cycle of the production process.

ILE4 developed computable general equilibrium electricity (CGE-E) model which is capable of performing scenario analyses on economic development. It is also capable of simulating the influence of exchange rates, taxation, and other policies on GDP, tertiary production growth, and national and regional industrial electricity supply and demand.

Agents Response Equilibrium Model (ARE)

Agent-based policy simulation is a new research method recently developed by domestic and international scholars. ARE is a policy-simulation model where a macro-system can be broken into micro-systems according to function, organization, or structure. Function abstraction is carried out on every micro-system. The Agent is a factor which controls them, and which contains a certain level of intelligence and reaction ability. Computers can simulate Agent interactions, decision making, and functions. As behavior changes at the micro level, various phenomena occur at the macro level. Based on fundamental macroeconomic principles, ARE model is price-oriented.

Based on principles of economics, one nation (region) could be set up in five kinds of Market Agents (MA): Commodity (products) Market Agent, Labor Market Agent, Financial Market Agent, International Commodity (products) Market Agent, and International Financial Market Agent; three kinds of Economic Operation Agents (EOAs): Enterprise (production) Agent (EA), Resident (consumption) Agent (RA), and Bank (Commercial Bank) Agent (BA). Each EOA is composed of lots of agents; that is, in EA class, there are thousands of agents producing heterogeneous commodities. It could be one hundred of steel-producing enterprises in EA to supply identical products and compete in market. In RA class, there are thousands of RAs featured as different incomes, consumption mentality, and life styles. In BA class, it could be dozens of BAs who have diverse business ideas and patterns and compete in market; there are also two types of Regulation Agents: Government Agent and Central Bank Agent. All these agents and their relations constitute the operation chart of economic system, as shown in Figure 2.7.

The market agent's primary function is to simulate marketplace mechanisms. Then, based on the supply–demand equilibriums, it adjusts the price of goods according to price adjustment rules. Therefore, its structure is relatively simple, and it primarily consists of five parts: identification, objective, strategic decision generation,

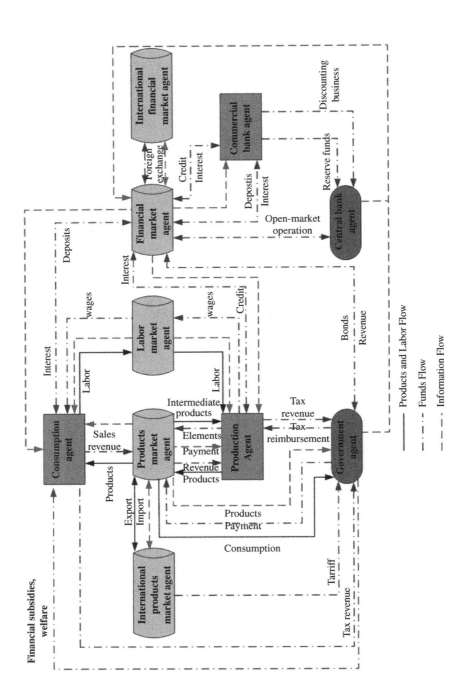

Figure 2.7 ARE model.

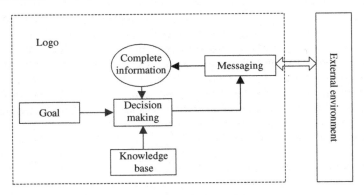

Figure 2.8 Market agent structure.

information communication, and a knowledge base. The structure is shown in Figure 2.8. The market agent receives its output and demand data from several factors. These include the enterprise agent, resident consumption demand, and government consumption demand. Prices are adjusted based on the supply–demand equilibrium.

The goal of the market agent is to create a market environment which is based on competition and enterprise supply. It actively adjusts market prices based on the product supply–demand equilibrium, and these adjustments are promptly fed back to the industry agent. The specific price adjustment rules are as follows:

1. If $D_i(t) < Y_i(t)$ then $P_i(t) = P_i(t-1) + \lambda \dfrac{D_i(t) - Y_i(t)}{D_i(t)} p_i(t-1);$

2. If $D_i(t) \geq Y_i(t)$ then $P_i(t) = P_i(t-1) + \lambda \dfrac{D_i(t) - Y_i(t)}{D_i(t)} p_i(t-1)$

Y_i represents the aggregate supply, D_i represents the aggregate demand, and λ is a constant coefficient.

The industry agent is responsible for production. It pursues profit and organizes production and market sales based on the factor and production markets. The agent's production decision process completely simulates a real industry's decision process. First, based on accumulated knowledge and experience, and considering the anticipated market, the agent plans the production scale for the next production period. It also decides on fixed asset investments, labor input, and intermediate input. The agent obtains capital from the bank and labor from the labor market. The industry agent then sends its output, and demand for other industry products, to the market agent. The market agent determines prices based on the supply–demand equilibriums. The industry agent's goal is to achieve the highest profit possible according to definite decision rules. It adjusts production rules and goes through a finite number of cycles until the entire economic system reaches supply–demand equilibrium.

With the description of the production decision-making process as provided by the enterprise agent, the complex process is then divided into eight parts. The goal of the enterprise agent is to maximize its own profits. It retains information on the status of the last action (growth or construction production scale), the results of the action (changes in benefit), and the product market supply–demand equilibrium. Within the knowledge base, this reasoning is made up of two parts: the antecedent and the succedent. The antecedent uses the agent's internal status to determine what will happen in the next cycle (increased or decreased production). It then calculates the production result, including its own change in profit and the changes in the production supply–demand equilibrium. The succedent adaptively adjusts production rules based on benefit variations and the supply–demand equilibrium. The industry agent's output adjustment rules are as follows:

1. If (Benefit = 0 and Sign = 0 and Balance = 0) then Action = $-aS$;
2. If (Benefit = 0 and Sign = 0 and Balance = 1) then Action = bS;
3. If (Benefit = 0 and Sign = 1 and Balance = 0) then Action = $-aS$;
4. If (Benefit = 0 and Sign = 1 and Balance = 1) then Action = bS;
5. If (Benefit = 1 and Sign = 0 and Balance = 0) then Action = $-aS$;
6. If (Benefit = 1 and Sign = 0 and Balance = 1) then Action = $-aS$;
7. If (Benefit = 1 and Sign = 1 and Balance = 0) then Action = bS;
8. If (Benefit = 1 and Sign = 1 and Balance = 1) then Action = bS.

For the above, benefit represents the changes in profit brought about by the last action. 0 indicates the reduced profit and 1 indicates the increased profit. Sign represents the state of the last action. 0 indicates the decreased production and 1 indicates the increased production. Balance represents the market supply–demand equilibrium after the last action. 0 indicates that supply exceeded demand and 1 indicates that supply was less than demand. a and b are constants that can be used to change the scope of production.

Under social production, residents possess the factors of production. Resident agents sell these factors of production in order to obtain remuneration. This remuneration ultimately ends up in two primary places: resident savings and resident consumption. In order to simulate influences of changes to resident consumption levels, as well as economic structure and electricity consumption, we can simplify consumer behavior using one resident agent to represent the entire community's behavior. We can then use that to design an open agent. The open agent's consumption levels and structure are exogenous variables. While the whole system is experiencing supply–demand equilibrium, it can examine every product's added value and electricity consumption. It can also set up scenarios, conduct experimental comparisons, and analyze economic operations and electricity consumption under the different scenarios.

In the macroeconomic system, economic system controllers are even more important than the consumers. The goals of the government agent are economic stability, growth, and social efficiency. Its primary action is to regulate economic system performance by way of macroeconomic policies. The government's source of income is taxation. Its expenditures include two parts. The first part is transfer payments, which ultimately return to the residents. Residents then use these payments for consumption or deposit them in the bank as savings. The second part is the money that governments spend on public construction. Fiscal policy mainly influences macroeconomic performance by adjustments to tax rates and changes in government spending.

Therefore, monetary policy develops bank credit by adjusting bank interest rates and reserve ratios. The policy also influences macroeconomic performance through public market activity, as well as the buying and selling of national debt. The government agent's control measures have been simplified and it is viewed as an open agent. Tax rates and government spending levels and structure are exogenous variables. They are set up by the researcher. Due to the open agent design, the government agent's goals and actions can be modified by the researcher. Then, researcher decides on tax rates, as well as government spending levels and structure based on a synthesis of knowledge and their own judgment. Once the simulation parameters are set up, the simulation is run. The results are then examined. Based on these results, adjustments are made to relevant policies to allow for different comparisons.

Apparently, the model is dynamic. It is able to depict the decision-making behaviors of all economic men (agents) to the market information, specifically, e.g., how do consumers (RA) decide the commodities amount to purchase by reacting to commodities' prices and self ability to pay, and how do they choose the saving amount based on self-demand and financial market information; how do producers (EA) plan the production amount based on commodities' prices information and production costs; how do banks (BA) determine credit amount via market information. These agents' responses (decision-making) to market information become an operation process of economic system or an iteration. After multiple times of such economic operation process (multiple times of iteration) and finally reaching balance, the whole system will have experienced the entire process of economic operation. This process could be recorded in detailed data which is useful for us to analyze and study. The ARE model is capable of the following functions: (i) it could depict the supply and demand curves of each commodity; (ii) it could depict the price curve of each commodity; (iii) it could depict the output curve of each commodity; (iv) it could depict the value added of each commodity; (v) it could simulate the fiscal policy effect; (vi) it could simulate the monetary policy effect; (vii) it could generate an Input–Output Table at each economic operation (iteration); (viii) it could simulate and examine on some economic theories or viewpoints.

Figure 2.9 shows the impact of some monetary policy on GDP. Figure 2.10 shows the influence of some monetary policies on the prices of the 42 sectors.

While CGE can give several results after the economic system reaches equilibrium, ARE can dynamically display every iteration (week) of the process, including prices, output, output value, demand etc., for all goods. Thus, the entire dynamic economic process is displayed weekly, and it is easy to see the consequences of certain policies. This is one of the greatest advantages of using ARE model.

2.3.2 Electricity Supply–Demand Forecasting Models

There are three main types of electricity supply–demand forecasting model.

LEAP Model

LEAP (long-range energy alternative planning system) model was developed through a collaborative effort between Stockholm's Environmental Research

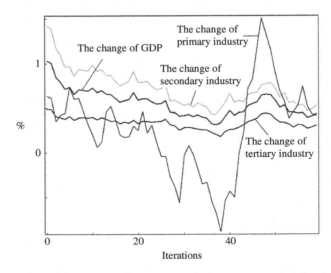

Figure 2.9 Monetary policy and the effects on the whole economy and tertiary industry.

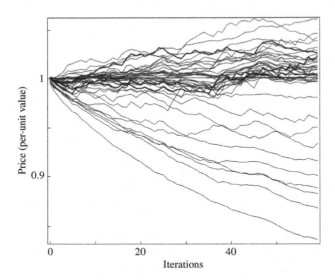

Figure 2.10 Monetary policy and the effects on 42 sectoral prices.

Institute and Boston University in the United States. It is bottom-up energy–environment model. The internal technology and environment database (TED) can be used for computing energy demand and the resulting pollutant emissions.

LEAP model is made up of two modules: the end-use demand module and the energy conversion module. The end-use demand module computes every sector's demand on every power source. This demand is based on the given activity levels

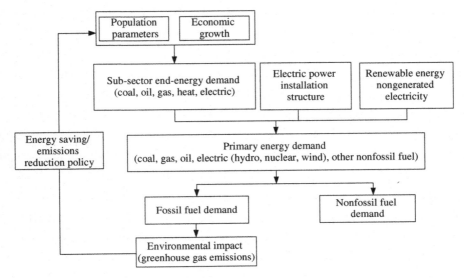

Figure 2.11 Basic principles of LEAP modeling.

of each sector (added value, production output, or service volume), and the corresponding type and quantity of energy demand. In LEAP model, the end-use demand module can carry out a detailed analysis of end-use energy technology by entering data on specific energy equipment technology. The module can also analyze energy demand trends based on the input of macroeconomic parameters.

LEAP combines "bottom-up" and "top-down" methods to carry out model building. Its basic principles are shown in Figure 2.11.

Using "top-down" methods, future economic development and industry segment added value are determined. Using statistical analysis on the growth of industry added value, growth in primary industry production output can be determined. By the comparison of domestic and international industrial energy and electricity intensity, the future industrial energy and electricity intensity can be calculated. Then the LEAP model can simulate the final energy demand for different sectors under different production level and energy efficiency (including coal, coke, oil, natural gas, heat and electricity), combined with the process conversion efficiencies of electrical and thermal energy, the total demand of primary energy can be calculated.

We deducted the nonfossil energy which did not used for generating electricity from the primary energy demand and then comprehensively considered the changing trends of the demand structures of primary energy to finally get the demand for each primary energy, such as coal, oil, natural gas, and electricity (hydro, nuclear, and wind). The LEAP model tree structure is shown in Figure 2.12.

The primary industry includes agriculture, forestry, livestock, and the fishing industry. In addition to planting and animal husbandry, this also includes agricultural produce processing, services, as well as drainage and irrigation. Industry added value

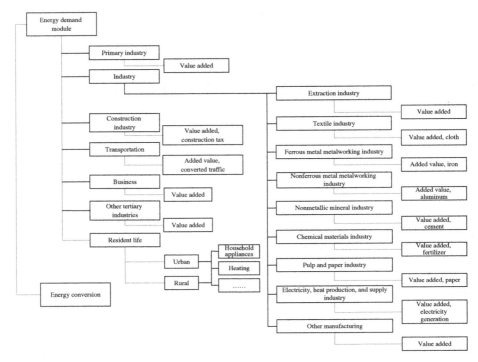

Figure 2.12 LEAP model tree structure.

acts as production activity indicator. The demand of different types of oil will take the total growth of agricultural machinery power into account. Mining includes the coal mining and purification industries, oil and natural gas extraction, ferrous-metal ore mining, mining of nonferrous metals, mining of nonmetal ores, and other mining industries. It includes every type of mineral resource mining, purification, and production sorting activity. Industry added value is a characteristic indicator of production activity.

The textile industry produces all types of yarn, cloth, and synthetic fibers. Industry added value and fabric output are characteristic indicators of production activity.

The main products of the ferrous-metal smelting and pressing industry are iron, steel, and iron alloys. The production scale and the production process of these products have a comparatively large influence on energy demand. Iron output typically acts as a characteristic indicator of production activity, and added value acts as a secondary check.

The main products of the nonferrous-metal metalworking industry are aluminum, copper, zinc, and rare earth elements. Aluminum output and energy demand levels are comparatively high. For this reason, aluminum output is used as a characteristic indicator of production activity, and added value acts as a supplementary check.

Nonmetal mineral production includes cement and flat glass. Energy demand is comparatively high for cement production, and, for this reason, cement output is used as a characteristic indicator of production activity. Here, added value also acts as a supplementary check.

The manufacturing of chemical materials and products includes soda ash, caustic soda, ethylene, and fertilizers. Fertilizer production uses a comparatively large proportion of energy and electricity. Therefore, added value and fertilizer output are characteristic indicators of production activity.

The pulp and paper industry produces all types of paper and cardboard. Paper occupies a comparatively large portion of production and is therefore used as the characteristic indicator of production activity. Here, added value also acts as a secondary check.

In the electrical power and heat power production and supply industry, electrical energy production is the characteristic indicator.

For other manufacturing industries, trades, and additional tertiary industry, added value is used as the characteristic indicator.

In the construction industry, the primary characteristic indicator is annual constructed area, and added value acts as supplementary check.

Residential is divided into urban and rural residents. The primary considerations here are per capita housing area, domestic electric appliance use, heating, etc.

In energy processing and conversion, heat supply and electrical power production are the main numbers of interest. Under electrical power production, the main consideration is adjustments to the structure of electricity sources.

Sectoral Analysis Model

Sectoral analysis model conduct sector-specific electricity demand forecasting. This is based on national economic industry divisions and industry segment classifications. Then, a model is created in order to obtain total electricity demand.

Sectoral analysis model carry out forecast analyses based on the electricity demand of the primary, secondary, and tertiary industry, as well as residential demand. The forecasts about electricity demand by the three industries are done by way of regression analyses or by energy intensity method. Urban/rural residential electricity demand is predicted by regression analyses or by per capita electricity demand. The basic principle of the sector analysis model is shown in Figure 2.13.

A decision analysis can be carried out on the future energy demand of each industry. These are based on the special relationship between the tertiary industry electricity demand and the economy, changing trend of per unit energy intensity and progression of technology, etc. By combining the forecasts of three industries incremented value, forecast future electricity demand can be made.

Establishing the regression equation of urban/rural residential electricity demand to the total population, urbanization rate and other indicators to forecast the urban/rural residential electricity demand based on the analysis of future population and urbanization changes. Corresponding influences on total electricity demand can be obtained from electricity demand by the tertiary industry and urban/rural households.

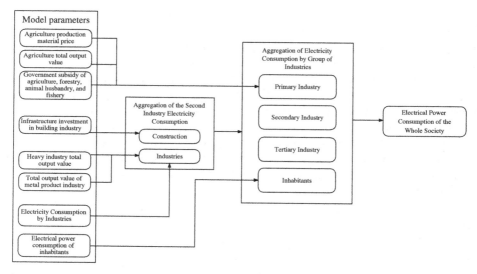

Figure 2.13 Basic principles of sectoral analysis model.

Input–Output Model

Using input–output (IO) model, mid- and long-term electricity demand can be forecasted. Based on future economic growth, electricity demand is then determined for each sector. Leveraging this model to demonstrate the marketing economy of "production according to the demand" which is a very interpretive "demand-oriented and production meets demand" concept, from where we can clearly explain the proportion of electricity demand. The use of input–output model to simulate mid- and long-term electricity demand is shown in Figure 2.14.

Based on an economic scenario analysis, every industry's total production can be calculated. Furthermore, total end-use vectors of each industry can also be determined.
Based on total industry production, every industry's electricity demand is determined.
Based on consumption structure decomposition, residential electricity demand is determined.
Based on residential electricity demand and each industry's electricity demand, total electricity demand is calculated.

2.4   The Scenario Analysis Method of Economy and Electricity Demand from ILE4

In ILE4, scenario analysis is used to study China's mid- and long-term economy and electricity demand. First of all, based on the analysis of China's economy development in the past and the determination of factors which might affect the future, we applied the CGE model to simulate the future economy growth of China and its different regions. Second, based on the economy analysis of China and its different

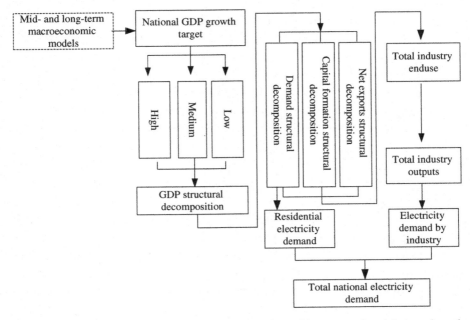

Figure 2.14 Using input–output model to simulate mid- and long-term electricity supply and demand.

regions, and application of the energy and electricity demand analysis and forecast model of ILE4, we simulated China's overall energy demand, electricity demand, and regional electricity demand. Comparison methods are then used to make the national and regional forecasting results identical. This scenario analysis system (and related ideas) is shown in Figure 2.15.

1. Decisions are made based on several future conditions. These include international economic and political environments, the domestic economic development environment, urbanization, employment structures, investment, consumption, exports, and technological developments. Corresponding exogenous variables are set up, and CGE model is used to conduct a scenario on the composition of the national and regional (region, city) mid- and long-term economy and industry. Exogenous variables include the population and size of labor force, total factor productivity, the balance of payments, tax rates, urbanization levels, global GDP growth rate, resident savings rates, government expenditures, and the composition of fixed asset investment.

2. Based on the results of the economic scenario analysis, energy-using technology developments, social production, and residential electricity consumption efficiency, LEAP model is used to forecast national energy and electricity demand. They can then obtain results for the following: primary energy demand by subvariety (coal, oil, natural gas, nuclear power, hydroelectric power, wind power, biomass energy generation, etc.), total energy demand, end-energy demand, and subsector subvariety end-energy demand (including electricity demand), etc.

3. Based on the economic scenario analysis of each provincial (region, city), utilizing the regression model, sector analysis model, production per unit consumption model and

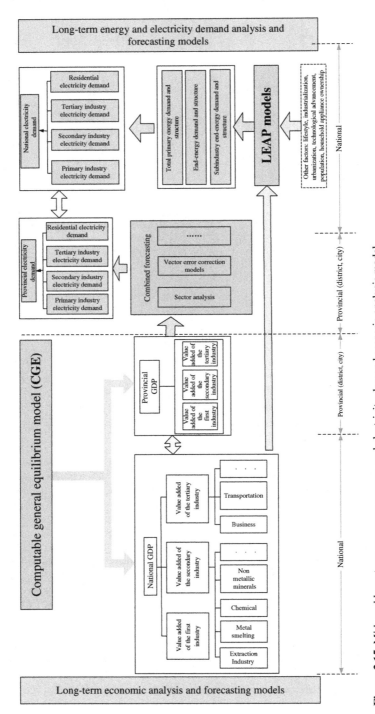

Figure 2.15 Mid- and long-term economy, energy, and electricity demand scenario analysis model.

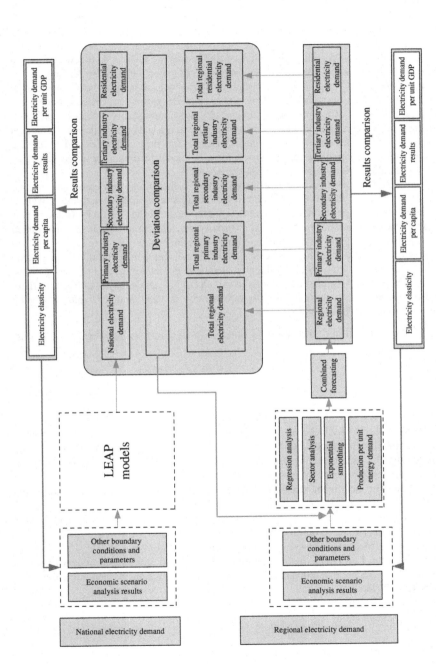

Figure 2.16 Electricity demand scenario analysis.

exponential smoothing model, etc., as well as the integrated forecast model to obtain the pattern of electricity demand growth and its structure in the mid- and long term.

4. "Bottom-up" method is used to summarize total provincial (region, city) and regional electricity demand. Based on the differences between this value and the predicted national value, the national and regional electricity demand results are revised. In this way, regional electricity demand can be determined as a proportion of national electricity demand.

5. Using the "top-down" method, decomposition is carried out on regional electricity demand as a proportion of national demand. The result is provincial (region, city) electricity demand.

6. Based on the results of the provincial (region, city) economic scenario analysis, and considering future industry development, sector analysis model and electricity intensity model is used to forecast and obtain the provincial (region, city) three industries and residential electricity demand; applying the integrated forecast model to obtain the three industries and residential electricity demand after the integration.

7. The demand of the provincial (region, city) tertiary industry and residential are summarized and compared. Based on provincial (region, city) and national deviations, as well as electricity composition trends, adjustments are then made to the provincial (region, city) electricity demand composition. Finally, electricity demand compositions are determined for the provincial (region, city) tertiary industry and residential.

8. Based on provincial (region, city) electricity demand and composition, the provincial (region, city) tertiary industry and residential electricity demand is calculated. Based on the results of provincial (region, city) economic and electricity demand forecasting, electricity intensity, the electricity elasticity, per capita electricity demand, and per capita residential electricity usage can be computed.

The method of national and regional electricity demand scenario analysis is portrayed in Figure 2.16.

Notes

[1] Zhaoguang Hu, et al. Electricity supply-demand simulation research—soft science laboratory of intelligent engineering. Beijing: China Electric Power Publishing House; 2009.

[2] Zhaoguang Hu. Intelligent engineering—its application. Proceedings of the IEEE international conference on systems, man and cybernetics; 1995;1:609–14.

[3] Zhaoguang Hu. Intelligent Modeling. Asia Energy Environment Modeling Forum: How to Model low Carbon Economy in Asia. Beijing University; 2007.

[4] Xu Minjie. Intelligent engineering and its application in electrical power supply and demand analysis and forecasting [doctoral thesis]. Beijing: Beijing Jiaotong University; 2008.

[5] Qi Huanqing. Macroeconomics. Beijing: Qinghua University Press; 2007.

[6] Zhang Yunqun, Lou Feng. Analysis and prediction of mid and long term growth potential in China's economy: 2008–2020, Quantitative and Technical Economics Research 2009;12:137–45.

[7] Tan Xiandong. Computable General Equilibrium Electricity model construction and applied research. 2008.

3 A Review of China's Economic Development and Electricity Consumption

3.1 Review of China's Economic Development

3.1.1 Economic Growth

After experiencing economic reform and opening up, China's economy has maintained a high rate of growth. China's GDP grew from 1.5 trillion Yuan in 1978 to 31.4 trillion in 2010 (in constant 2005 Yuan), with an average annual growth rate that reached 9.9%. Although China's economic development has experienced several low points, (such as the 1998 Asian financial crisis and the 2008 global financial crisis), China's economy has been able to advance in a smooth and stable manner. This has been due to the fact that it has benefited from the direction of a number of counter-crisis measures. In particular, during a 7-year period from 2001–2007, China experienced its longest span of rising economic development since the reforms, with mostly stable economic performance. Overall national strength and international competitive ability increased substantially. This helped to lay a solid foundation which allowed China's economy to catch up with the economies of rest of the world. In 2006, China's GDP exceeded Britain's. In 2007, it exceeded Germany's, and, in 2010, it exceeded Japan's, becoming an economic powerhouse second only to the United States.

Despite the fact that its economy currently ranks second in the world, China's economic development remains comparatively low when viewed from the standpoint of average GDP. According to current exchange rates, China's GDP per capita in 1978 was 155 USD. By 2010, it had reached 4379 USD, a 28-fold increase. However, in the world rankings, China's GDP per capita is still quite low, standing in approximately 95th place.[1] In the 1960s, the United Nations defined one characteristic of a "modernized country" as one where the GDP per capita has reached 3000 USD. According to calculations made by China's Academy of Social Sciences,[2] if USD inflation and deflation are considered, this figure would be the equivalent of a modern-day value of 8000–10,000 USD. China's GDP per capita has reached 4000 USD, and therefore, it is still classified as a "mid-low income country."

An Exploration into China's Economic Development and Electricity Demand by the Year 2050.
DOI: http://dx.doi.org/10.1016/B978-0-12-420159-0.00003-5

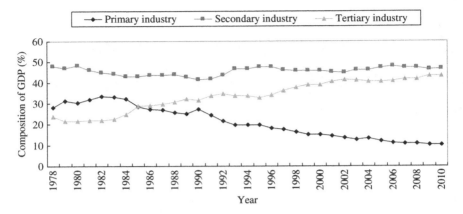

Figure 3.1 Structural changes to China's GDP from 1978 to 2010.

3.1.2 Industry Structure

Changes in China's economic structure from 1978 to 2010 are shown in Figure 3.1. Some characteristics of these changes are presented below:

In China, the primary industry, as a proportion of the overall national GDP, has steadily decreased. During the initial stages of the economic reforms, the primary industry's added value made up 30% of the national GDP. However, in 2010, this percentage had already fallen to 10.2%, a massive degree of reduction. It is also important to note that, from the initial stages of economic reforms through the mid-1980s, the proportion of GDP taken up by the primary industry's added value appeared to experience an upward trend. However, from the mid-1980s and onward, this trend reversed itself and headed downward. Furthermore, after the beginning of the 1990s, this decline was even more significant. The increases seen in the primary industry before the mid-1980s may be related to the national promotion of the household responsibility system, which strengthened agricultural production significantly at that time.

The secondary industry's proportion of the GDP initially dropped and then rose, but the total changes were not significant. As a percentage of the GDP, the secondary industry decreased from 47.9% in 1978 to 41.34% in 1990. In 2010, the secondary industry rose again to 46.8% of the GDP. Overall, the secondary industry's added value has occupied the most important place in the structure of the GDP. It has not, however, experienced major changes in percentage of the GDP.

The tertiary industry, as a proportion of GDP, has experienced its own increases. After 1983, the tertiary industry grew rapidly, eventually exceeding the primary industry by the year 1985. In 2002, the disparity between the tertiary industry and the secondary industry was very small, with a difference of only 3.07 percentage points. However, since 2002, the percentage of the GDP occupied by the tertiary industry's added value has changed slowly.

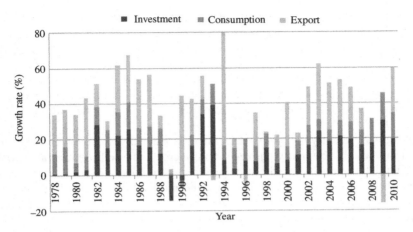

Figure 3.2 Real growth rate of investment, consumption and exports since economic reform.

3.1.3 Investment, Consumption, and Exports

Overall investment in China has been steadily expanding. Though the investment growth rate has fluctuated to some extent, total growth remains relatively high, and it is the main driving force behind economic growth. As shown in Figure 3.2, from the outset of economic reforms through the 1990s, China primarily encouraged basic investment. During this period, fixed asset investments showed comparatively rapid increases, which in turn helped to spur China's speedy economic development forward with an average annual investment growth rate of 12.7%. From 1990 to 2000, during the reform period, China's fixed asset investments intensified. Furthermore, control was obtained over the investment structure by adjusting the tax rates on various industry investment items. In 2000, the total capital formation contribution to the economic growth rate was 22.4%. Post-2000, China's investment rate has been maintained at comparatively higher levels and has been the dominant force behind economic growth. Due to China's continued industrialization and urbanization, and combined with rapid economic development, the annual growth rate of fixed asset investments has been increasing. Between 2000 and 2010, it averaged 26.8%.

Fluctuation in consumer demand has been relatively modest, and it has progressively become a new driving force behind economic growth. It remains a relatively stable economic performance factor and has demonstrated both high stability and low volatility. Between 1978 and 1990, China's economic development played an important role by helping to solve the population's need for adequate food and clothing (Figure 3.2). The consumption growth rate continued to rise rapidly, with an annual average of 10.3%. From 1990 to 2000, the most important justification behind rapid economic development was to help increase the standard of living. During these 10 years, consumption growth began to decrease, with an annual average growth rate of 9.8%. Since 2000, consumption levels have increased once again, alongside resident incomes and consumer confidence. There has also been a growth in the total retail sales of consumer goods. Housing, automobiles, home computers, and other advanced electronics equipment have become popular consumer goods in China. Despite the global financial crisis, consumption has been able to maintain its relatively rapid growth, and it has progressively become a new driving force behind overall economic growth. Over this period of time, China's total consumption levels have experienced comparatively large increases, reaching an annual average growth rate of 11.3%.

Along with its entry into the WTO, China has gradually become the world's manufacturing base. The country has seen rapid increases in both imports and exports. Foreign trade has gradually gained more and more influence over the economy, and exports have increasingly become one of the most important factors driving economic development. From 1978 to 1990, under the national "Export for Foreign Reserves" policy, China's export growth rate was able to maintain its relatively high increases. This remained true for all but 1 year during this period (Figure 3.2). During this time, the average annual growth rate was 30.1%. From 1990 to 2000, China's export growth rate remained stable, with a low average annual growth rate of 24.0%. Over the past few years, largely due to China's increased competitive prowess, as well as anticipated yuan appreciation, the value of imports and exports has rapidly increased alongside a continuously expanding trade surplus. Since 2008, due to the global financial crisis, imports and exports have shown marked decreases, and their contribution to GDP has been lowered significantly. Net exports boosted the GDP growth rate by a mere 1.2 percentage points. This phase has had an annual average growth rate of 20.6%.

3.1.4   Industrialization

Over the past 30 years, the factors that have served to advance China's industrialization processes have been aided by the optimization of industry structures. Additionally, China's industrialization process is now showing continual and regular improvements. Not only has GDP per capita surpassed 1000 USD, but the composite index of industrialization now exceeds 33.3.[3] In early 2000, China entered a period characterized by heavy industrialization. This process continued into the second half of 2005 (by then, the composite index of industrialization had already reached 50). In 2010, the composite index of industrialization increased further, rising to about 68.

Nationwide, the industrialization index (Table 3.1) appears to be geographically distributed. It progressively decreases from east to west. The eastern region is relatively developed with high levels of industrialization. In contrast, the industrialization process continues to lag in the western regions, with almost all areas still in the beginning or middle stages of industrialization. Over the past few years, industrialization has made rapid advances, but Tibet still remains in the primary stage of industrialization. A large discrepancy exists between Tibet and the other provinces.

Regionally, the eastern areas of Beijing and Tianjin have already entered postindustrialization. Shandong is approaching complete industrialization and Hebei, Shanxi, and Inner Mongolia are all in the latter midterm stages of industrialization.

Table 3.1 Nationwide and Provincial (District, City) Composite Indexes of Industrialization

Region	1995	2000	2005	Region	1995	2000	2005
National	18	26	50	Sichuan	4	8	25
Beijing	81	92	>100	Chongqing		15	34
Tianjin	73	83	96	Liaoning	38	43	63
Hebei	14	24	38	Jilin	16	24	39
Shanxi	16	22	45	Heilongjiang	22	33	37
Shandong	20	33	66	Shaanxi	10	14	30
Inner Mongolia	5	13	39	Gansu	15	11	22
Shanghai	89	99.5	>100	Qinghai	7	15	30
Jiangsu	34	45	78	Ningxia	11	15	34
Zhejiang	32	47	79	Xinjiang	9	17	26
Anhui	4	7	26	Tibet			
Fujian	21	35	56	Guangdong	35	55	83
Henan	6	11	28	Guangxi	2	4	19
Hubei	13	28	38	Yunnan	15	13	21
Hunan	2	11	28	Guizhou	4	6	13
Jiangxi	2	8	26	Hainan	6	10	17

Source: Academy of Social Sciences official industrialization report, "2007 industrialization report," "China industrialization report—1995–2005 China's provincial industrialization assessment and research."

Due to use of the phased threshold method, it was not possible to measure growth for regions in the post-industrial stage. Regions with a composite index of industrialization of 100 are regarded as completely industrialized and in the post-industrialization phase.

Chongqing was established in 1997 and therefore 1995 had no index. The deflator is around 8.

At present, Tibet is in the pre-industrial stage and therefore has no index.

Shanghai has already entered the post-industrialization stage, and Jiangsu and Zhejiang are approaching complete industrialization. Fujian is in the first half of late industrialization, and Anhui is still in the first half of mid-industrialization. In the middle regions of the county, the total levels of industrialization are relatively similar. All areas are in the midterm stages of industrialization. Among these, Hubei and Chongqing have entered the second half while the remaining four provinces are in the first half. Each of these regions remains at a lower level than the average nationwide industrialization level. Liaoning, in the northeastern region, has entered the final stages of industrialization, while Jilin and Heilongjiang are still in the second half of mid-industrialization. Guangdong is approaching complete industrialization. Hainan, Guangxi, Yunnan, and Guizhou are still in the first half of industrialization.

3.1.5 Urbanization

Along with the gradual reduction in the barriers between urban and rural society, China's urbanization has advanced rapidly. This is especially true for the development of smaller towns. The urbanization rate has continued to rise, and this has further spurred China's rapid economic development. In 2010, it was around 47.4%, an increase of 30 percentage points since 1978. China's urbanization is divided into two important phases. These are demonstrated in Figure 3.3. The first phase occurred from 1978 to 1991 when urbanization was rapidly developing. Due to the increase in urban construction under the economic reforms, the urbanization rate increased from 17.9% to 26.9%. The second phase began in 1992 and has continued through to the present. China's urbanization is now in its mature stages. By using the development of China's coastal areas as a representative example, we can see how the nation has now entered into stable urbanization. China's urbanization rate in 1978 was 17.9%, a figure which had increased to 47.4% by 2010. This is a significant gain of nearly 30 percentage points.

Along with the steady increase in urbanization, China's nonagricultural population has also experienced some major changes. By examining the 2010 urbanization index for urban and rural populations, we can see that China's urban population now

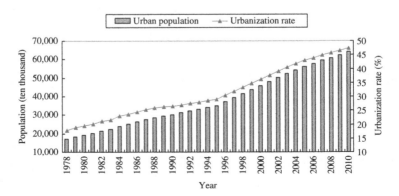

Figure 3.3 Urbanization rate and urban population since economic reform.

makes up 47.4% of the total. Within this population, the number of people employed in the nonagricultural sector exceeds 40%. According to the current proportion of the urban population and urbanization measurement standards, China is now in the midterm stage of urbanization. Furthermore, China's urbanization is shifting from the rapid to the mature phase of development.

3.1.6 Regional Development

China has experienced offset regional development due to differences in geographical locations, natural resources, and several other factors. The unbalanced nature of China's regional economic development is significantly more obvious. It shows a high degree of total regional economic disparity. An appropriate degree of regional disparity in economic development can have a positive effect, but when the inter-regional economic imbalance continuously increases, it can have negative implications for macroeconomic development.

From 1978 to 2010, the economic growth rate in the eastern regions had an annual average of 12.4%. This was 2.5 percentage points higher than the national average. The economy in the eastern region has experienced increases in the proportion of the national economy that it occupies, and has grown from 43.6% in 1978, to 52.7% in 2010. Conversely, the central and western regions have begun to occupy an increasingly smaller proportion of the total economy, which is a trend that has been especially pronounced since 1990. From 1978 to 2010, the GDP in the central region had an annual growth rate of 10.9%. This accounted for 20.1% of the national economy. When compared with 1978, this figure showed a decrease of 1.5 percentage points. The GDP in the western region had an average annual growth rate of 10.7%, and accounted for 18.6% of the total economy. This was a 2.2 percentage point decrease. The average annual growth of the GDP in the northeastern region reached 9.9%. Nationally, the economy in the northeast is comparatively small. In 2010, it only accounted for only about 8.6% of the entire national economy, and had decreased 5.4 percentage points since 1978. The regional economic growth rates and their proportions of the national economy from 1980 to 2010 are shown in Figures 3.4 and 3.5.

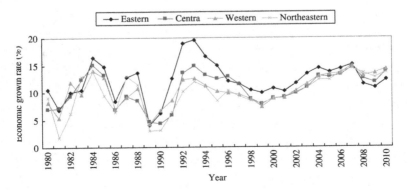

Figure 3.4 1980–2010 regional economic growth rate.

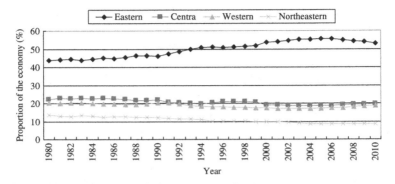

Figure 3.5 1980–2010 regional economic proportions of the national economy.

3.2 National Electricity Consumption

3.2.1 Total Electricity Consumption

In 2010, China's total electricity consumption reached 4.19 trillion kWh[4]. This was 16.8 times higher than the usage in 1978, with an annual average growth rate of 9.2% (Figure 3.6). Total electricity consumption in China increased from 1 trillion kWh to 2 trillion kWh over an 8-year period from 1997 to 2004. Consumption continued to rise from 2 trillion kWh to 3 trillion kWh during the next 4 years (2004–2007). The trend also continued from 2007 to 2010 when electricity consumption rose from 3 trillion kWh to 4 trillion kWh. Electricity consumption per capita has reached 3125 kWh, which is 12 times the amount used in 1978. This means that the average annual growth rate amounted to 8.1%.

At the beginning of the economic reforms, China gradually withdrew from its commitment to the simple development of heavy and light industry. This decision

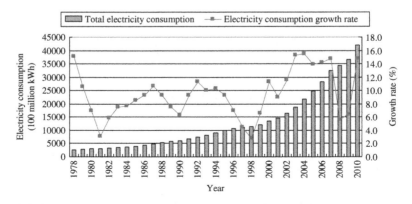

Figure 3.6 Growth in total national electricity consumption since 1978.

was partly responsible for the nation's ability to develop quickly. In the 1980s, China's annual average electricity consumption growth rate was 7.6%. In the 1990s, under the influence of the high-speed growth in the processing industry, electricity consumption experienced relatively high increases. During the overexpansion of the economy, there was national tightening of regulatory policies, and, in 1997, China suffered once again from the impact of the Asian financial crisis. However, in 1990, the annual average electricity consumption growth rate still reached 8.2%. At the outset of the twenty-first century, alongside the accelerated industrialization and urbanization process, the national economy emerged from the shadow of the Asian financial crisis and began to experience high-speed growth. Total societal electricity demand continued to increase. Throughout the beginning of the twenty-first century, despite the influence of the Asian financial crisis, the 10-year annual average electricity consumption growth rate reached 12.0%. Furthermore, it demonstrated a continual trend of accelerated development.

According to IEA statistical analysis, China's electricity consumption[5] was sixth in the world in 1980. Along with economic development, electricity consumption has also been increasing quickly. China has gradually advanced positions in the international rankings. In 1990, China's electricity consumption exceeded both Canada's and Germany's, moving up to fourth place. In 2000, China rose to second place. During the 2001 through 2009, despite remaining second in the rankings, China continued to occupy an increasingly large proportion of the world's electricity consumption. The electricity consumption[6] of several major countries is compared in Figure 3.7.

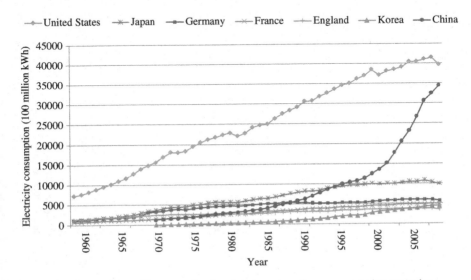

Figure 3.7 Comparison of increases in electricity consumption among major countries. *Source*: OECD iLibrary and China Electricity Council (CEC).

3.2.2 Electricity Consumption Per Capita

Along with its economic development, China's electricity consumption per capita levels have maintained their rapid increases and appear to be accelerating in development. From 1990 to 2000, due to macroeconomic policies[7] which emphasized an economic "Soft Landing," as well as structural adjustments, China's electricity consumption per capita increased relatively little. From 1990 to 2000, China's electricity consumption per capita increased from 536 kWh to 997 kWh, with an average annual growth rate of 7.1%. This figure was 1.1 percentage points less than the growth rate for total electricity consumption. In 2001, per capita consumption began to rise above 1000 kWh. It continued to increase, and had reached 3125 kWh by 2010. This was an average annual increase of 11.4% from 2001 to 2010.

A corresponding rise in the GDP per capita can be seen after 2003 when China had a GDP of approximately 10,000 Yuan (Figure 3.8). China's electricity consumption also continued to increase significantly. This period can be viewed as an accelerated growth period which arose after the initial economic takeoff. Economic increases are dependent variables of large increases in electricity demand.

Along with the improvements in living conditions, the popularity and consumption rates of household electrical appliances gradually increased. Residential electricity consumption per capita levels also showed some very large increases. From 1990 to 2000, residential electricity consumption per capita rose from 40 kWh to 132 kWh, with an average annual growth rate of 12.6%. After 2001, alongside the increasing popularity of computers, air conditioners, washing machines, and other electronic appliances, residential electricity consumption continued to maintain its high rate of growth. In 2010, electricity consumption per capita reached 382 kWh. From 2001 to 2010, the average annual growth rate was 11.2%. Table 3.2 shows the possession rate of electricity consumption appliance per 100 Chinese households each year.

When compared with other major developed countries, China's electricity consumption per capita is still very low (Figure 3.9). In 2009, China's electricity

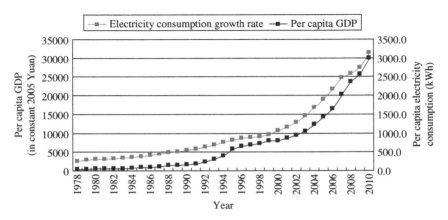

Figure 3.8 China's GDP per capita and electricity consumption since 1978.

Table 3.2 The Possession Rate of Electricity Consumption Appliance Per 100 Chinese Households

Item	Urban					
	1990	**1995**	**2000**	**2005**	**2008**	**2009**
Washing machines	78.41	88.97	90.5	95.51	94.65	96.01
Refrigerators	42.33	66.22	80.1	90.72	93.63	95.35
Color TVs	59.04	89.79	116.6	134.8	132.89	135.65
Stereo systems (set)		10.52	22.2	28.79	27.43	28.21
Air conditioners	0.34	8.09	30.8	80.67	100.28	106.84
Showers		30.05	49.1	72.65	80.65	83.39
Home computers			9.7	41.52	59.26	65.74

Item	Rural					
	1990	**1995**	**2000**	**2005**	**2008**	**2009**
Washing machines	9.12	16.9	28.58	40.2	49.11	53.14
Refrigerators	1.22	5.15	12.31	20.1	30.19	37.11
Color TVs	4.72	16.92	48.74	84.08	99.22	108.94
Air conditioners		0.18	1.32	6.4	9.82	12.23
Home computers			0.47	2.1	5.36	7.46

Source: "2010 China Statistical Yearbook"

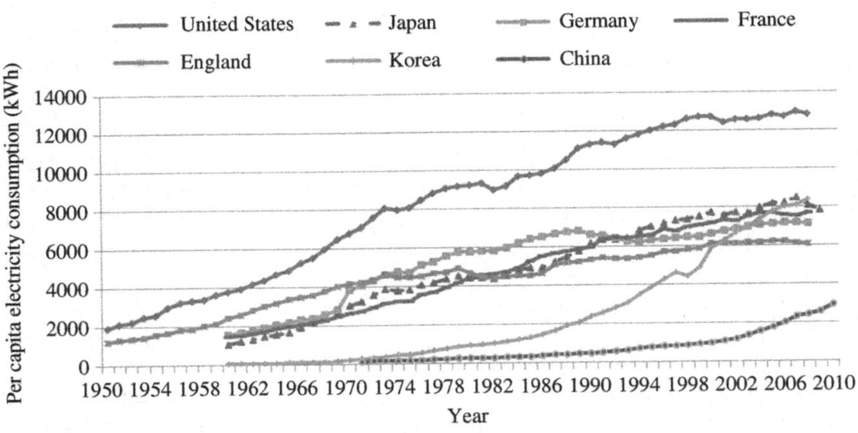

Figure 3.9 Comparison of increases in electricity consumption per capita for several major countries.
Source: OECD iLibrary and CEC.

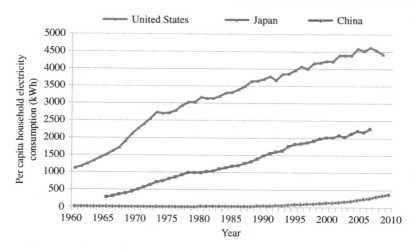

Figure 3.10 Comparison of increases in residential electricity consumption per capita for the major countries.
Source: OECD iLibrary and CEC.

consumption per capita barely reached one-fifth that of the United States, and it was about one-third that of Korea, Japan, France, and Germany. Currently, China's electricity consumption per capita is only equivalent to the levels in the United States during the 1950s (in 1955, 2961 kWh). It is also the equivalent of Japan's levels in the 1970s (1970, 3017 kWh), and Korea's during the early 1990s (1993, 3101 kWh).

When compared with developing countries, China's current electricity consumption per capita is approximately equivalent to Brazil's, but it is greater than those of India and Egypt.

The gap between China's residential electricity consumption per capita and the levels of other developed countries is even larger (Figure 3.10). In 2009, China's residential electricity consumption per capita was one-tenth of the level in the United States. It was one-fifth that of Japan's and France's, and one-third that of Korea's. By comparing China with some developing counties, we can also see some significant disparities. In 2009, China's residential electricity consumption per capita was just over half the consumption in Egypt and Brazil. The overall level only remained higher than that of India.

3.2.3 Electricity Consumption Structure

The increases in electricity consumption[8] by sectors are shown in Figure 3.11.

The electricity consumption by the primary industry has been affected by the fluctuating influences of air temperature and precipitation. From 1991 to 2000, the average annual growth rate was 5.7%. Since 2001, the average annual growth rate has reached 6.3%, which is lower than the total electricity consumption growth rate over the same period.

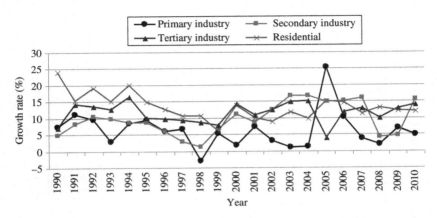

Figure 3.11 China's electricity consumption growth rate by sector since 1990.

The growth rate for secondary industry electricity consumption is a major factor which determines whether the total electricity consumption growth rate is higher or lower. From 1990 to 2000, structural adjustments caused the industrial electricity consumption growth rate to slow down. This was especially true for the industries with high-energy intensity. The average annual growth rate for secondary industry electricity consumption was 7.2%, which was slightly lower than total electricity consumption levels. Since 2000, the secondary industry electricity consumption growth rate sped up. This increase came alongside increases in both the industrial economy and the rapid development of industries with high-energy intensity. Despite the impacts of the global financial crisis from 2008 to 2009, the growth rate for secondary industry electricity consumption rebounded to a comparatively high level. From 2001 to 2010, the average annual growth rate for secondary industry electricity consumption was 12.3%. This, in turn, drove up total electricity consumption and helped to increase growth.

The tertiary industry electricity consumption was able to maintain a stable level of rapid growth due to a number of factors, including the accelerated development of the service industry. From 1991 to 2000, the average annual growth rate was 11.6%. Since 2001, the average annual growth rate has gone up to 11.8%, and has demonstrated relatively stable growth.

Residential electricity consumption[9] has also maintained relatively high growth rates due to the influence of improvements in living conditions and an increase in the availability of electricity. Since 1990, consumer supply and demand patterns have been radically altered. Color TVs, refrigerators, and other conventional household electric appliances were able to maintain their relatively high-speed growth rates. Air conditioners, computers, and communications equipment also gradually began to enter ordinary households. The average annual growth rate for residential electricity consumption was 13.7%. Since 2000, automobile and real estate consumption have also escalated. Residential electricity consumption continues to maintain a high growth rate with an average annual growth rate of 11.9%.

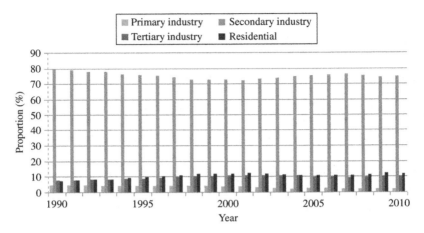

Figure 3.12 Proportion of electricity consumption in China since 1990.

Along with China's rapid economic development and improvements to living conditions, industry structure has continuously undergone adjustments and optimization. The electricity consumption structure has also seen its own share of changes. China's sub-industry electricity consumption structure (post 1990)[10] is shown in Figure 3.12. In 1990, China's three industries and residential electricity consumption composition was: 5.0:79.4:8.1:7.5. Soon, under the influence of technological improvements and structural adjustments, the proportion of electricity consumption by secondary industry began to decrease. At the same time, both the three industries and the residential electricity consumption rapidly increased under the improved living conditions and increased consumption levels. They gradually began to make up an increasing proportion of total electricity consumption. The proportion of electricity consumption made up by primary industry is relatively small and has steadily been decreasing. In 2000, China's three industries and residential electricity consumption changed to: 4.0:72.7:10.9:12.4. At the outset of the twenty-first century, heavy industrialization became a more and more pronounced as a part of China's economic development. Electricity consumption of secondary industry has had a high growth rate, and its proportion of total consumption has continuously increased. Conversely, the proportion of consumption occupied by primary industry has quickly decreased. Three industries and residential electricity consumption proportions have also decreased. In 2010, the three industries and the residential electricity consumption ratios were about 2.4:74.7:10.7:12.2.

The proportion of consumption occupied by the electricity consumption of secondary industry is no more than 40% in OECD countries[11] (Figure 3.13). The proportion of electricity consumption occupied by tertiary industry and residential is about 30%. Electricity consumption in primary industry is also extremely low. The proportion of electricity consumption of secondary industry in China is higher than other foreign countries. The proportions of electricity consumption held by both the tertiary industry and residential are significantly lower.

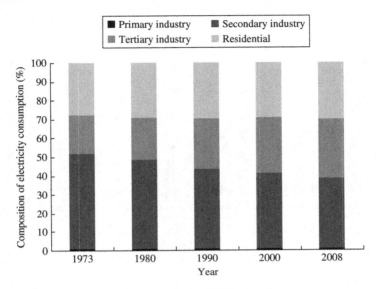

Figure 3.13 Electricity consumption structure in OECD countries.
Source: International Energy Agency (IEA) Yearly Electrical Information and Data.

3.2.4 Electricity Consumption in Key Industry Sectors

The four largest energy consumers in China are the ferrous metal, nonferrous metal, chemical, and nonmetallic mineral industries.

Since 1990, because of macro-control influences, the development of energy-intensive industries was restricted. The total electricity consumption growth rate was lower than the electricity consumption levels of society as a whole. The proportion of electricity consumption by energy-intensive industries decreased from 29.96% in 1990, to 26.9% in 2000. Since 2001, the four largest energy-intensive industries had an average annual growth rate of 14.0% for electricity consumption. This exceeded the total electricity consumption growth rate over the same period by 2 percentage points. In 2010,[12] electricity consumption for the four largest energy-intensive industries increased to 32.4%. The changes in the electricity consumption proportions for the four largest energy-intensive industries are shown in Figure 3.14. Electricity consumption by the ferrous and nonferrous metal industries experienced annual increases in their proportion of total consumption. Conversely, the electricity consumption by the chemical materials industry showed annual decreases in the proportion of total consumption. The building materials industry showed stable increases in its proportion of total electricity consumption.

The rapid development of energy-intensive industries has caused a trend toward heavy industry in terms of electricity consumption. In 1990, the electricity consumption structure for light and heavy industry was: 20.4:79.6. By 2000, this changed to: 20.8:9.2. In 2010, it was further changed to: 16.8:3.2. Since the turn of the century, China's heavy industry electricity consumption proportion has substantially increased.

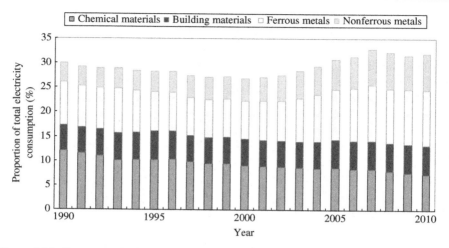

Figure 3.14 Changes in the proportion of total electricity consumption occupied by the four largest energy-intensive industries.

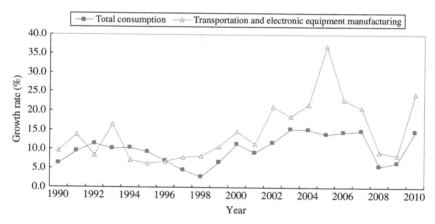

Figure 3.15 Electricity consumption growth rate for the transportation and electronic equipment manufacturing industries.

After 1990, structural adjustments for the transportation and electronic equipment manufacturing industries have allowed for comparatively rapid growth. Even though the fluctuations in industry electricity consumption are greater than those in total electricity consumption, the growth has still been faster than that of total electricity consumption (Figure 3.15). After 1999, the high growth rate of electricity consumption in the transportation and electronic equipment manufacturing industries caused the proportion of industry electricity consumption to increase significantly. In 2007, transportation and electronic equipment manufacturing became China's fifth largest industrial consumer of electricity.

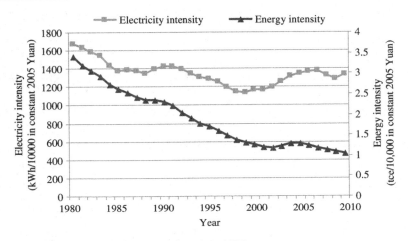

Figure 3.16 Electricity and energy intensity since 1980.
Source: Calculations published by the China Statistical Yearbook and CEC.

3.2.5 Electricity Intensity

Changes in electricity intensity[13] can be reflective of both technological advancement and structural economic adjustments. If we looked into different time period, we can see that the electricity intensity tends to decline gradually. During the period from economy reforming to 2000, we could notice that the electricity intensity showed a continuous decline except year 1982 and 1990. After 2000, under the driving force of high-speed industry development, especially energy-intensive industries, electricity intensity rose.

Due to the electricity intensity of secondary industry is normally around 6 times higher than that of both the primary industry and the tertiary industry. Since 2000, the electricity intensity of secondary industry increased. Similarly, the added value of the secondary industry made up an increased proportion of GDP. This caused electricity intensity to show a tendency to rise. In 2005, China's electricity intensity was 1340 kWh/10,000 Yuan (in constant 2005 Yuan). When compared with 2000, it increased by 180 kWh/10,000 Yuan. Due to the effects of both the global financial crisis and measures for energy savings and emissions reduction, China's electricity intensity was 1335 kWh/10,000 yuan during the year of 2010. During the "Eleventh Five-Year" period, this figure decreased by only 0.4%, which was far less than the energy intensity decrease of 19.1%. The changes in China's electricity and energy intensities, since 1980, are shown in Figure 3.16.

3.2.6 Electricity Elasticity

Since 1980, China's electricity elasticity has fluctuated, as can be seen in Figure 3.17. Due to the growth of heavy industry electricity consumption in the "Tenth Five-Year" plan, electricity elasticity was highest, at 1.33. The electricity elasticity

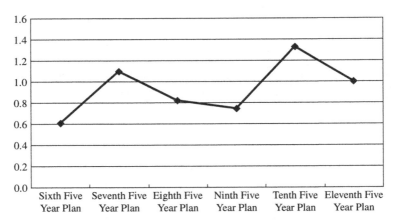

Figure 3.17 The changes of electricity coefficient in various stages since 1980.

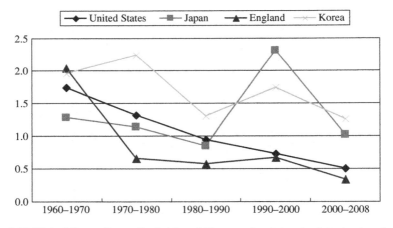

Figure 3.18 United States, Japan, England, and Korea—electricity elasticity by decade.

during the "Eleventh Five-Year" plan was slightly less than 1. More importantly, it should be noted that all of this occurred during the 2008 global financial crisis, which was responsible for substantial decreases in electricity consumption. From 2000 to 2010, China's average electricity elasticity was 1.15. This figure represents an increase over the 1980–2010 average, which was 0.92.

In periods of high-speed industrialization, most countries show a relatively large electricity elasticity. Generally, it is in the range of 1–2.5, as shown in Figure 3.18. After industrialization is completed, the electricity elasticity begins to recede. Under stable economic conditions with relatively slow growth rates, the electricity elasticity is less than 1. For example, in the United States, after World War II and continuing through the early 1970s, industrialization continued and the electricity elasticity was around 2.1. However, starting in the 1980s after industrialization had been completed, for every 10-year range the electricity elasticity has been lower than 1. Japan is one exception to this rule. In 1990, Japan's average annual economic growth rate was 1.18%, and total electricity consumption was driven by residential energy consumption. However, total electricity consumption still maintained an average annual growth rate of 2.52%. Thus, even with an extremely low economic growth rate, Japan's electricity elasticity exceeded 2.

Since 2000, China's electricity elasticity was also greater than 1. This coefficient remains closely related to China's economic development stage.

3.3 Regional Electricity Consumption

3.3.1 Regional Electricity Consumption

The eastern region of China was the first to experience economic reforms. Since 1980, the electricity consumption growth rate has usually been higher than total average level. Since 2003, largely due to increases in land and labor force costs, industries started moving toward the interior. This caused electricity consumption growth rates in the eastern areas to slow slightly. During the 30 years following 1980, the average annual electricity consumption growth rate was 10.16%, which exceeded the total average levels by 0.9 percentage points.

The economy of the central regions is not as developed as it is in coastal areas. Heavy industry development is also not as pronounced as it is throughout the rest of the country. Over the past few years, these disparities in development, found in the central regions, have increased due to reforms in the west. They are also due, in part, to a revival of the old industry base in the northeast, and more rapid development in the east. Neither geographical advantages nor policies have helped the situation. The result is what has been termed a "central collapse" phenomenon. Since 2000, along with implementation of a strategy for central emergence, the central region's electricity consumption has made relatively large increases. They have shown an annual average reaching 11.69%. This indicates a shrinking gap when compared with the national levels of consumption.

The economic foundation in China's western inland, border, and ethnic areas is quite poor. In the 1980s and 1990s, the electricity consumption growth rate here managed to hover around average national level. However, in recent years, along with the proposals for western development, and its plentiful natural resources, the economic growth rate in the west has gradually accelerated. This has helped to drive electricity consumption higher, as well as the corresponding increases in growth rates. Since 2000, along with development of energy-intensive industries, electricity consumption in the western region has witnessed comparatively large increases. It has shown an average annual growth rate that reached 13.29%, which was higher than average national levels by 1.26 percentage points.

The northeastern regions are China's old industry base. Over the years, the northeastern regions have faced economic structure conflicts. They have also

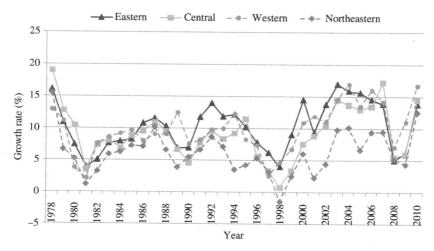

Figure 3.19 Regional growth of electricity consumption since 1978.

demonstrated a weak capacity for the sustainable development of natural resources. The growth rate here for total electricity consumption has consistently been lower than the national average.

Regional electricity consumption growth rates since 1978 are shown in Figure 3.19.

From 1980 to 2010, the proportion of electricity consumption occupied by the eastern region has grown substantially larger, although it has gradually been overtaken by western and central areas post-2003. In 2010, the proportion of electricity consumption held by the eastern region was about half of the national consumption. Compared with the 30 previous years, this was an increase of 11 percentage points. The eastern region's status as the energy consumption powerhouse has gradually been consolidated. The proportion of electricity consumption from the central regions declined until 2007, after which it finally began to increase. Since 2000, the proportion of national electricity consumption from the western region has increased significantly. The northeast has become the area with the lowest electricity consumption, and has shown an accumulated reduction of about 11.3 percentage points over the last 30 years. The regional proportions of electricity consumption since 1978 are shown in Figure 3.20.

3.3.2 Regional Electricity Consumption Per Capita

In the eastern region, electricity consumption per capita, and residential electricity consumption per capita levels are the highest in China. In 2009, eastern electricity consumption per capita was 3796 kWh, which was 12.8 times higher than in 1978. It also showed an average annual growth rate of 8.6%. This rate exceeded the national average by 0.7 percentage points over the same time period. Since 2000, the average annual growth rate has reached 10.79%, which is lower than the national

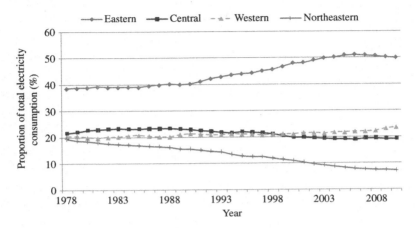

Figure 3.20 Regional proportion of electricity consumption since 1978.

average by 0.32 percentage points. Due to comparatively high resident income levels in the eastern region, the average residential electricity consumption is now well ahead of the national average. In 2009, residential electricity consumption per capita reached 541 kWh in the east. Since 1990, the annual average growth rate has reached 12.27%, exceeding the national average by 0.87 percentage points.

Electricity consumption per capita in the central region is on the lower side of the scale. In 2009, central electricity consumption per capita was 1982 kWh, which was 9.9 times higher than in 1978. The average annual growth rate was 7.7%, which was lower than national average by 0.2 percentage points. In 1978, electricity consumption per capita in the central region was lower than the national average by 61 kWh. By 2009, this gap had slowly expanded, reaching 760 kWh. In the central region, residential electricity consumption per capita was also lower than the national average. In 2009, residential electricity consumption per capita was 272 kWh in the central region, amounting to barely 74.7% of the national average.

Electricity consumption per capita in the western region is lower than the national average; however, it has been rapidly increasing since the year 2000. Total levels continue to be lower than the national level. In 2009, electricity consumption per capita was 2302 kWh in the western region, 12.4 times the 1978 level. It also demonstrated an average annual growth rate of 8.5%, higher than the national average over the same time period by 1.6 percentage points. From 2000 to 2009, the average annual growth rate was 12.65%, higher than the national average of the same period by 1.54 percentage points. Although the electricity consumption per capita growth rate in the western region was higher than the national average, economic lag has kept the absolute value below the national average. In 1978, the disparity between electricity consumption per capita in the western region and the national average was 75 kWh. This gap had grown to 440 kWh by 2009. Although the electricity consumption growth rate is comparatively high in the western region, and residential

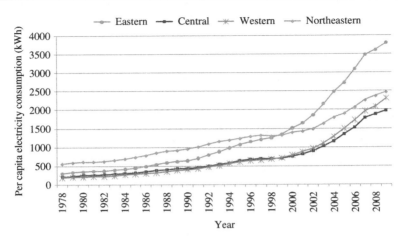

Figure 3.21 Regional electricity consumption per capita.

electricity consumption per capita has reached 271 kWh, it is still the lowest of all regions.

Electricity consumption per capita increases annually in the northeast, but the growth rate is significantly lower than the national average. In the 1980s and 1990s, electricity consumption per capita in the northeast was higher than the national average. However, due to slow increases in electricity consumption in the region after 2006, electricity consumption per capita was lower than the national average. In 2009, electricity consumption per capita was 2473 kWh in the northeast, lower than the national average by 269 kWh. In addition, residential electricity consumption per capita in the northeast has been high, but the growth rate is lower than national average. In 2009, residential electricity consumption per capita was 353 kWh in the eastern region, lower than the national average by 11 kWh.

The regional electricity consumption per capita and residential electricity consumption per capita are shown in Figures 3.21 and 3.22.

Amongst the provinces, electricity consumption per capita is highest in Ningxia, Qinghai, Shanghai, Inner Mongolia, and Zhejiang. These regions are all either western areas with strong concentrations of energy-intensive industries, or developed eastern coastal provinces. The regions with the lowest electricity consumption per capita are Hunan, Anhui, Hainan, Jiangxi, and Tibet. These are all central or western provinces, as shown in Figure 3.23.

The regions with the highest residential electricity consumption per capita are Shanghai, Beijing, Tianjin, Guangdong, and Fujian. All of these are eastern provinces, economically developed, and possess high standards of living. Areas with the lowest residential electricity consumption per capita include Shanxi, Jiangxi, Xinjiang, Tibet, and Gansu. All of these are central regions (Figure 3.24).

A comparison of the provinces shows that residential electricity consumption per capita better reflects economic development levels than electricity consumption per capita.

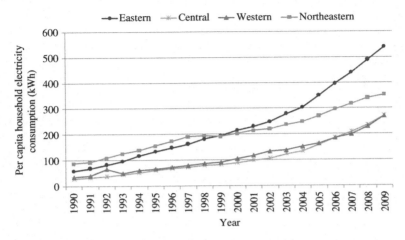

Figure 3.22 Regional residential electricity consumption per capita.
Source: Data from the National Bureau of Statistics and CEC.

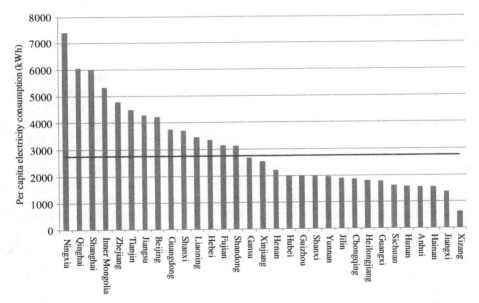

Figure 3.23 2009 provincial (district, city) electricity consumption per capita.
Source: Data from the National Bureau of Statistics and CEC.

3.3.3 Regional Electricity Consumption Structure

The proportion of electricity consumption for primary and secondary industry in the eastern region is significantly lower than the national average. Also, the proportion occupied by tertiary industry is higher than the national average. The proportion

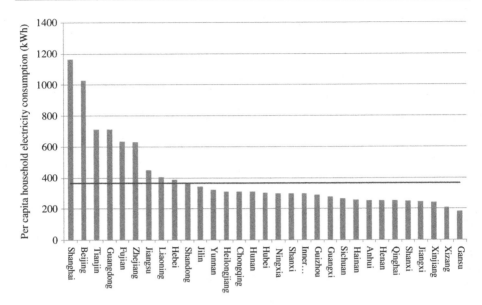

Figure 3.24 2009 provincial (district, city) residential electricity consumption per capita.
Source: Data from the National Bureau of Statistics and CEC.

of residential electricity consumption is close to the national average. For 2009, in the eastern region, the proportion of electricity consumption by three industries and residential consumption was 2.13%, 73.83%, 12.67%, and 12.37%. Since 1990, the proportion of electricity consumption by primary industry has steadily been on the decline. The electricity consumption proportion by secondary industry has maintained a fluctuation between 70% and 80%, first falling and then rising. Increases were relatively significant, especially since 2002. In 2007, once again it reached a peak level of 75.29%. In 2008, declines followed in the wake of the global financial crisis. The proportion of electricity consumption in the tertiary industry has shown a continuous and stable increase. In 1998 and 1999, the industry electricity consumption growth rate declined because of the Asian financial crisis. This allowed the proportion of electricity consumption occupied by residential to reach its highest point. The proportion of industry electricity consumption then increased, and the residential electricity consumption decreased, which corresponded with the economic recovery.

Primary industry, secondary industry, and residential electricity consumption proportions in the central region are all relatively high, while the proportion of tertiary industry electricity consumption is lower than the national level. In the central region for 2009, three industries and residential electricity consumption proportions were 3.47%, 73.82%, 9.07%, and 13.64%, respectively. Considering this trend, which can be observed since 1990, the proportion of electricity consumption by primary industry has been steadily declining in the central region. The proportion of secondary industry electricity consumption initially dropped and then rose, with particularly significant increases since 2002. After the impact of the financial crisis in 2008, there was a decline in electricity consumption. The proportion of electricity consumption occupied by the tertiary industry has increased steadily, while the proportion used by residential has increased rapidly.

Proportions of electricity consumption by primary and secondary industry in the western region are relatively high, but consumption by the three industries and residential show lower proportions than the national average. In the western region during 2009, the three industries and residential electricity consumption proportions were 2.95%, 77.49%, 8.15%, and 11.41%, respectively. Since 1990, electricity consumption by primary industry has steadily declined in the west. The proportion of electricity consumption in the secondary industry first decreased and then increased, with especially significant increases since 2002. Within the western region, the electricity consumption by the secondary industry in Qinghai and Ningxia exceeded 90%. The proportion of electricity consumption in tertiary industry first rose and then dropped, while the proportion of residential electricity consumption has shown mostly stable increases.

In the northeast, the proportion of electricity consumption by the tertiary industry is lower than the national average. However, residential electricity consumption is at a higher proportion than the national average. Due to the fact that it has a long history of being an industry base, the northeast occupies a comparatively high proportion of the national economy, with complete industrial divisions. A major proportion of the region's electricity consumption is used by secondary industry. Furthermore, because of geographical and climate factors, the northeast residential heating period is comparatively long. This is partially the reason why residential electricity consumption occupies a comparatively high proportion of the overall consumption in the region. In the northeast during 2009, the three industries and residential electricity consumption proportions were 1.97%, 73.60%, 10.55%, 13.88%, respectively. Electricity consumption proportions by the primary and secondary industry declined by 0.012% and 8.88%, respectively; when compared with 1990 levels, the consumption by the three industries and residential rose by 3.48 and 5.52 percentage points. The 2009 proportions of electricity consumption, both national and regional, are shown in Figure 3.25.

3.3.4 Regional Consumption of Electricity by Energy-Intensive Industries

Energy consumption growth rates by energy-intensive industries in the eastern region are slightly higher than the national average. Among these, ferrous metal working, chemical materials manufacturing, and building materials have shown the highest electricity consumption. From 1990 to 2009 in the east, the four largest energy-intensive industries had an average annual electricity consumption growth rate of 10.37%. This was higher than the national average by 0.26 percentage points, but lower than overall eastern consumption rates by 0.78 percentage points. This, in turn, meant that energy-intensive industries had a lower energy consumption proportion than the national average. After 1990, in the east, the proportion of electricity consumption of

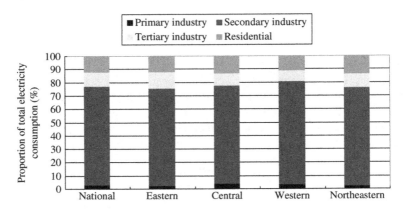

Figure 3.25 National and regional electricity consumption proportions in 2009.
Source: Data from CEC.

the four largest energy-intensive industries initially dropped and then began to rise. Since 2002, growth rates for electricity consumption have reached 20%, thus garnering a rapidly increasing proportion. In 2009, the four energy-intensive industries in the east had an electricity consumption proportion of 23.36%.

Energy-intensive industries in the central region have an electricity consumption proportion that is higher than the national average. Since 1990, these four industries had electricity consumption proportions which initially dropped and then rose. Since 2002, the electricity consumption growth rates have all been higher than 15%, gaining a rapidly increasing proportion of overall consumption. In 2008, due to the impacts of the global financial crisis, energy-intensive industries represented a decreased proportion of electricity consumption. In 2009, these industries in the central region took up 38.14% of consumption. When this figure is compared with 1990, it was an increase of 6.45 percentage points, and was higher than the national average by 6.52 percentage points. Amongst these four industries, the ferrous metal working industry accounted for 11.64% of the overall electricity consumption, which was higher than the national average by 4.6 percentage points. Due to these grandiose proportions since 2000, energy-intensive industries now account for about 40% of overall consumption in the central region.

Energy-intensive industries in the western region also show a high electricity consumption proportion. Here, these types of industries have shown faster development than the national average. Since 1990 in the west, these industries showed slight declines followed by rapid increases in their proportions of electricity consumption. Since 2002, the electricity consumption growth rates of the four industries have all been around 20%. This means that their proportions of consumption have increased rapidly. In 2007, energy-intensive industries amounted for 50.13% of total electricity consumption. However, after the impact of the financial crisis, these industries all saw their proportions of electricity consumption decrease. In 2009, the four major energy-intensive industries accounted for 45.73% of total consumption, an increase of 7.95 percentage points from 1990. Furthermore, this figure exceeded the national

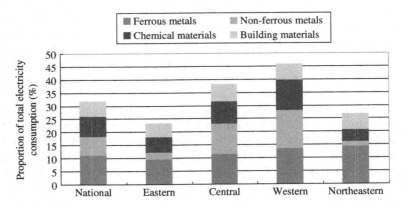

Figure 3.26 The proportion of electricity consumption of national and regional high-energy intensity industry in 2009.
Source: Data from CEC.

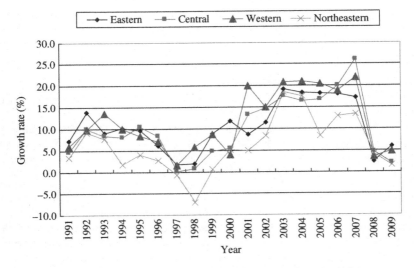

Figure 3.27 The regional electricity growth rate of high-energy intensity industry.
Source: Data from CEC.

average by 14.11 percentage points. Chemical manufacturing, ferrous metal, and the nonferrous metal industries accounted for more than 11% of the overall electricity consumption. Since the electricity consumption of the high-energy intensity industries is relatively high, these industries in Western region have accounted for approximately 50–60% of overall electricity consumption since the year 2000.

The development of energy-intensive industries has been relatively slow in the northeast. They, therefore, account for a lower proportion of consumption than the national average. In the northeast region's industrial system, the industries with the highest consumption demands are cement, steel, and iron alloys. These industries are mainly concentrated within three, large-scale, nationalized companies (SOEs) and are named as follows: Shenyang Heavy Industries, Anshan Steel, and Changchun Yatai Cement. Production at these facilities has been relatively stable, with small overall load fluctuations. Historically, however, industries in the northeast with high energy demands have been slow to develop. They have also held a much lower proportion of consumption than the national average. In 2009, these four types of industries consumed 71.9 billion kWh of electricity, accounting for 26.7% of the overall consumption (among these, ferrous metals comprised 14.16%, nonferrous metals comprised 1.98%, chemical manufacturing comprised 4.36%, and building materials comprised 6.20%). In this area, Industries with high energy demands accounted for a lower proportion of consumption than the national average by 4.92 percentage points.

The 2009 proportion of electricity consumption for national and regional industries with high-energy intensity is shown in Figure 3.26. The electricity consumption growth rate by region for industries with high-energy intensity demands is shown in Figure 3.27.

Notes

[1] Data from IMF, found at http://www.IMF.org.
[2] China Academy of Social Sciences. 2010 Official Report. Beijing: Social Sciences Academic Press; 2010.
[3] As proposed by the Chinese Academy of Social Sciences, the industrialization process is divided into five indexes: GDP per capita, tertiary industry proportion of output value, manufacturing industry proportion of added value, tertiary industry proportion of employment, and urbanization rate. The whole industrialization process is divided into the pre-industrialization stage (with the sum of the indexes = 0), early stages of industrialization (>0, <33), midterm (>33, <66), late (>66, <99), and post-industrialization (>100). The early, mid, and late stages of industrialization are also divided into first and second halves. Chen Jiagui, Huang Qunhui, Zhong Hongwu, Wang Yanzhong et al. China industrialization report (1995–2005 China provincial industrialization assessment and research). Social Sciences Documentation Publishing House; July 2007.
[4] Bulletin released by the China Electricity Council (CEC) January 2011.
[5] There are inconsistencies within the data released by the CEC. According to IEA statistics, electricity consumption = total electricity generation + import electricity − export electricity − line power loss.
[6] IEA statistics. Electricity Information 2010, International Energy Agency.
[7] Wang Chunzheng. Exploration practice: macroeconomic performance and control. Beijing: Economic Science Press; 2003.
[8] Zhaoguang Hu, Baoguo Shan, Xinyang Han et al. China electricity demand outlook—based on electricity supply-demand laboratory simulations. Beijing: China Electric Power Press; 2010.

[9]   Shi Dan. The new electricity consumption situation: non-industrial electricity consumption increases. Chinese Power Industry, 2000, 1.

[10]   Since 1986, China's electricity consumption classification structure has experienced comparatively large changes. This data is not comparable with data collected prior to 1985. For a convenient analysis, this book gives a detailed breakdown of the data beginning in 1990.

[11]   Energy balance of OECD countries, IEA, 2010.

[12]   2010 industry forecasting data.

[13]   State Grid Energy Research Institute. Power supply and demand analysis report (2010). Beijing: China Electric Power Press; 2010.

4 Analysis of Major Factors Which Affect Electricity Demands

4.1 The Growth of the World's Economy

Since the recession of 2008, international economics has entered a new phase. Over the long term, the world's economy will even out, and return to fostering growth. From the perspective of economic development, crisis and structure adjustments are associated with the birth of a new system for increasing the production sector,[1] and to help bring the economy out of recession into a new era.

First, from the perspective of the integration of international economics, although trade protection and exchange rate disputes are currently on the rise, economic globalization will continue to be a major trend in the future, and it will continue to promote growth.

Overall, the future of international economic development will have the following characteristics:

First, the international division of labor will expand further and continue to entrench itself. The international industrial sector will expand to an even greater scale. The international division of labor will appear in a more intricate pattern, and, not only will the vertical division of labor be intertwined with the horizontal division of labor, but the refinement of the split based on the product chain and the industrial chain products within the division of labor will also continue to develop. This will help to promote the expansion of trade products within the industry to push forward further growth in trade, services, and investment. This will eventually create even more international trade and investment opportunities.

Corresponding to economic globalization, the regional economic integration process will also be greatly accelerated. Over recent years, various forms of regional economic cooperation and free trade have been flourishing vigorously; intraregional trade and investment are strong stimuli for the development of international trade and direct investment. From this trend of development, regional economic cooperation and regional economic integration will continue to deepen, and products, capital, technology, personnel, and rates of mobility in the region will be more frequent. This is because there is still much room for further development of intraregional trade and investment.

Second, according to the global economic pattern, the status of the major emerging economies will continue to rise.

In the process of economic restructuring, the international market location structure will change, and the importance of the emerging economies on the market will continue to increase. From the perspective of economic growth, developed countries

An Exploration into China's Economic Development and Electricity Demand by the Year 2050.
DOI: http://dx.doi.org/10.1016/B978-0-12-420159-0.00004-7

will face a higher proportion of public debt, economic recovery, and employment recovery. There will be further potential risks for economic downturn. Therefore, the economic growth in developed countries may be slower. By contrast, emerging markets and developing countries will continue to promote their industrialization and urbanization, and grow into the peak construction period. This will, not only generate demand for goods and services, but also generate investment and various types of infrastructure, public services, and facilities. This greater demand for infrastructure, public services, and facilities is expected to maintain rapid economic growth. Even though the developed countries might seek to hinder the growth rate in the emerging markets and developing countries, their growth rates will still be higher than the global average. They will only be further enhanced and continue their contributions to global economic growth. From a product demand point of view, due to the slow recovery of demand in developed countries, (particularly the smaller growth elasticity of low-end labor-intensive products, electromechanical products, and high-technology products), the growth elasticity of capital and technology-intensive products are reflected amongst developing countries. Overall, the right of the major emerging economies to be heard, and to influence the global economy in terms of international economic and financial affairs will be significantly enhanced. Therefore, the form the pattern of the world economic order will be undergoing profound changes.

Again, in terms of industrial development, resources, and environmental constraints, as well as new energy sources, energy saving technology and other green industries, will lead to future global industrial transformation.

According to the theory of economic growth, economic structural adjustment in the process of "Creative Destruction" is the basis on which to drive the economy back into the growth channel. Every large-scale innovation will be made out of older technology and production systems, and these will help to establish a new kind of production system. The international financial crisis, which was triggered by a global economic structure adjustment, will also provide development opportunities for the emerging industrial sector.

From the perspective of a long-term fundamental driving force to promote world economic growth, economic globalization, science, and technological progress will continue to promote growth. Intensive efforts made by countries to develop new energy sources, energy saving technologies, and other green industries may trigger a new wave of scientific and technological progress. They may also cause the formation of a new economic growth point. This effort is becoming the new driving force which is seeking to promote world economic growth after the internet. Of course, a number of related industries will also gradually become more mature and play a role, not only in driving the upstream and downstream development of the industry and the formation of large-scale industrial clusters, but also in promoting the transformation and upgrading of traditional industries. This will then trigger a new global economic structure adjustment and reorganization. These industries may also help to spur a new round of economic growth.

In summary, the international financial crisis, and subsequent structural adjustment, will not affect the long-term growth trend of the global economy. According to the forecasts from the World Bank, IMF, the International Energy Agency, and other international institutions, over the next two decades, the global economy will continue growing at a speed of 3~4%. Furthermore, the global scale of economies will grow from 35 trillion USD in 2005, to about 72 trillion USD in 2030 (based on constant 2005 market exchange rates and constant prices).

4.2 Domestic Economic Restructuring

The international economic situation will affect China's economic growth. Domestic energy conservation and emissions reduction policies will also play a role. China's future economic growth rate may begin to slow moderately, especially while economic development focuses on economic restructuring and changes in the mode of development.

Industrialization and urbanization are the important driving forces behind industrial restructuring. Currently, China is in the mid and late stages of industrialization. The process of industrialization is now showing accelerated development. Future promotions of the industrialization process, as well as industrial modernization, are still important tasks for China's economic development. Industrial growth from the main volume expansion will gradually be shifted toward quality improvement. According to the Composite Index of China's industrialization trends, China is expected to complete its industrialization around 2020. Also, because the levels of industrialization amongst China's various regions are not balanced, advancing industrialization is a difficult task for the regions between the gradient. As a developing country, China has the opportunity to achieve a great-leap-forward in its development by both learning from the achievements of other advanced countries. It can then try a newer and more enlightened road toward full industrialization.

The speedy development of strategic emerging industries will be first on the agenda for China's industrialization. The official release of the state council on the decision to speed up the cultivation and development of the strategic emerging industries includes plans for energy savings, new sources of energy, vehicles powered by nontraditional energies, new materials, biomedicine, high-end equipment manufacturing, and a new generation of IT industries. These seven strategically emerging industries will be the country's future focus while it seeks to support the industrial sector. In the future process of industrialization, the heavy chemical industry will continue to maintain its rapid pace of development. The development of the heavy chemical industry is the inevitable result of the evolution of the economic structure in the process of global industrialization. The United States, Japan, and other developed countries have experienced a process of industrialization that has gone through the stage of rapid development for the heavy and chemical industry. Its leading industries are all basically the same; they are mainly automotive, machinery, iron, steel, petrochemical, and others. The rapid development of the heavy and chemical industry is essential for China to complete its industrialization process. China is a country in which the industrialization process has already reached a significant level. The trend should be that energy-intensive industries remain in stable development, but the direction should be compatible with the global realities of environmental protection, energy savings, and environmental production. Recently, ferrous metals, nonferrous metals, materials for construction, and chemical industries with high energy consumption have continued to show great potential for growth. However, with the advancement of the industrialization process, the production of products within these industries will gradually become saturated. After complete industrialization, the growth of relatively energy-intensive industries will slow. This will create new leading industries that will focus on strategic emerging parts of the economy. Therefore, the automotive, pharmaceutical, electronics, and electrical equipment industries will become more prominent, while the importance of iron and steel, nonferrous metals, and other heavy industries will be gradually reduced.

Similar to industrialization, China's urbanization is seeking not simply to increase the proportions of the urban population but also to promote the coordinated development of both urban and regional urbanization. According to the three stages of the "S" shaped curve of urbanization development, China's urbanization level is currently in the second phase of accelerated development. The government is in the post-crisis era used to promote the process of transformation of the mode of economic development. Therefore, the government will seek to increase the promotion of industrialization and urbanization. It has been estimated that, before the year 2020, China's urbanization rate will witness an average increases of 0.85~1.2 percentage points. After 2020, the rate of urbanization increase will begin to decline significantly, and the average annual increase would be about 0.6~0.9 percentage points. It is expected that, in 2020, the urbanization rate will range in the area of 56~60%, and by 2030, it will be around 61~68%. The development of urbanization will not only need the support of the steel and cement industries but will also require a change in people's habits and energy usage. It will ultimately promote growth in the overall demand for electricity. According to the historical data from 1978 to 2009, China's urbanization rate increased 1.6 times, and accordingly, the electricity consumption per capita increased 9.6 times. Furthermore, residential electricity consumption per capita has increased 23.5 times.

Urban agglomeration will be the main method of China's future urbanization development. The period before 2020 will see a phase of initial development and the positioning of the regional urban agglomerations. According to the regional urban layout, and the level of economic development, in the future China will see the formation of a group of 10 major cities, namely, Beijing, Tianjin, the Yangtze River Delta, Pearl River Delta, Shandong Peninsula, Liaoning Central South Central Plains, the middle reaches of the Yangtze River, the west side Sichuan, Chongqing, and other urban agglomerations. Currently, most city groups are still demonstrating the preliminary stages of development, support policies, estate planning. Functional positioning is still in the exploratory stages. These characteristics are not included for the Pearl River Delta, Yangtze River Delta, and the Beijing-Tianjin-Hebei group, which are more developed major cities. Over the next 10 years, under the established national urbanization strategy, these city groups will have made great progress. The years between 2012 and 2030 will be an integration period for the major urban agglomerations. The top 10 urban agglomerations are set to be the anchors of the future, in-depth development, in China's urbanization strategy. Not only do they cover China from the east coast to the southwest, through the vast areas of the northwest hinterland, but the development of the top 10 urban agglomerations will also have a profound impact on China's overall economic layout. City development through the regional industry cluster, resource optimization, and a rational division of labor will be able to complement one another. There should be coordinated development and an expansion of economic growth, as well as the balancing of disparities among regions. All of this will help to speed up the integration of the domestic market, and to promote economic development.

The advance of industrialization and urbanization will directly contribute to industrial restructuring. In China, there are three main sectors, agriculture, industry, and services. On the one hand, the third industry's value added will change significantly. According to national planning and historical trends, it is expected that the agricultural added value, in terms of the proportion of GDP, will continue to decline through the year 2020. Furthermore, the industrial added value percentage of the GDP will also decline significantly. The Economic and Social Development of the People's Republic of China's Twelfth Five Year Plan has tried to promote the development of the services sector, as well as economic structure optimization. The upgrading of the strategic focus of the future of our value-added services share of GDP will rise significantly. Around the year 2030, the service industry will continue to increase significantly. On the other hand, the adjustment of economic structure is not only reflected in the changes in GDP, and, more importantly, the tertiary industry's production method and operation method will be significantly improved.

Before 2020, in the agriculture sector, China will continue to build new socialist rural construction projects, as well as work on the development of this critical period for modern agriculture. However, China will also try to break the dual structure of urban and rural integrated development during this the crucial period. Then, after 2020, modern agriculture will be extended. Not only will modern production methods be promoted, but also new operational modes of agriculture, industrial chain layouts, and the regional division of labor will further be optimized and integrated with rural production. The result of all of this will be that the quality of citizens' lives will be greatly improved. The supply of electricity for both rural-residential and industrial needs will continue to grow and help to create a new socialist countryside.

In the industry sector, 2011–2030 will be an important period for the completion of China's industrialization, upgrading, and improvement. Before 2020, China will promote industrialization in order to speed up the stage of completion. This will be characterized by a transitioning from heavy and chemical focused industries, to an industry which is focused mainly on processing. The industry sector will gradually move from general or resource-intensive processing industries, to technology-intensive processing industries. This will speed up the process of upgrading the economic structure. The slow growth of traditional industries will cause the status of the entire industrial system to drop.

In the services sector, during the period from 2011 to 2030, China's traditional services will gradually begin to transfer to a more modern service industry. In the period before the year 2020, with the steady progression of the industry sector, production services, especially research and development, finance, insurance, logistics, creative, and information will show greater development. The process of circulation in these areas will also improve and increase growth. At the same time, catering, leisure, and other traditional kinds of services will continue to play a labor-intensive role and help to absorb employment. Therefore, during this period, there will appear to be a steady development in the traditional services industry, while the modern services sector will also appear to grow rapidly. After 2020, not only will the service industry become dominant over the entire national economic system, but the modern services industry will also be the dominant force for new support systems such as intelligent transportation systems, intercity transport networks, e-money applications, transaction security, network services, and the credit system. All of this will greatly improve the technical level of the service industry. Service quality and operational efficiency will fully reflect modern information and knowledge-intensive features. The proportion of service sector using electricity will continue to rise steadily, but then growth will begin to slow.

In addition, the regional economic layout adjustment will show some generalized changes in terms of economic structure, but, in view of its importance, and how it is rich in content, we will develop this topic further in Section 4.4.

4.3 The Development of Energy-Intensive Industries

According to the history of development patterns in the world's major developed countries, during the stage of industrialization, a country tends to show rapid growth in terms of mineral resources consumption. At this stage, metallurgical, construction, and chemicals, as well as production and the primary processing of energy-intensive industries grow very rapidly. These are also the industries that need large amounts of raw materials. This leads to a high metallic mineral resource consumption growth rate which is generally greater than, or equivalent to, the GDP growth rate. Alongside this, source production and consumption per capita also rise. As the industrialization completes, the infrastructure is gradually completed. This will significantly slow the development of energy-intensive industries. The per capita resource production and consumption growth rates are also gradually reduced, and the result can be negative growth.

By examining the development of the iron and steel industries in the United States, Japan, Britain, Germany, and other developed countries, crude steel production and consumption per capita in the 1970s reached a record high (Figure 4.1 and Table 4.1). When the steel consumption per capita reached its peak in these countries, the proportion of the industry was in a declining phase. They had had basically completed their industrialization and the urbanization rates had reached more than 65%.

According to the historical development path followed by the developed countries, China's steel industry still has much room for development. From the perspective of stages of economic development, China is still in the industrialization and urbanization accelerated development period. As long as urbanization and industrialization are not at the end of their growth period, the demand for steel will also not end. The steel industry will continue to maintain long-term development trend. From

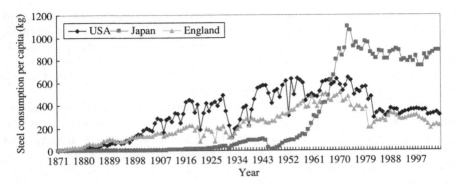

Figure 4.1 Steel consumption per capita of the United States, Japan, and England.
Source: Japan Iron and Steel Union Steel Statistics Committee: "Iron and Steel Statistics View" (1970) World Steel Association: steel Yearbook (2000 Edition and 2010 Edition) Angus Maddison: The World Economy A Millennial Perspective". Prior to 1950, population data is derived according to the "World Economic Millennium History of the projections."

Table 4.1 Some of the Industrialized Countries, Peaks in Crude
Steel Consumption Per Capita

Countries	Consumption Peak (kg)	Occurrence Time (year)
USA	711	1973
Japan	802	1973
England	473	1964
Germany	660	1970

Source: Xu, XiangChun. Investigation into the peak of China's steel consumption. http://finance.sina.com.cn/review/observe/20070307/18583385510.shtml.

the steel consumption per capita in 2009, China's steel consumption per capita is about 420 kg, which is equivalent to the United States' consumption levels during the 1940s, and to Japan's levels during the 1960s.

According to the analysis of China's industrialization process, China as a whole will gradually complete industrialization somewhere between the years 2010–2020. The industrialization sector will gradually shift from a rise to a stable downward trend, and there will be a gradual slowing of growth in steel production. We are expecting that a saturation point for iron and steel production is likely to appear around 2020. By taking into account the fact that advances in technology, and the development of alternative products, will help reduce the demand for steel, this seems to be the most likely result. Along with the levels of upstream energy and resource constraints, China's steel industry has reached a production peak. Steel consumption per capita will become lower than the United States and Japan, and settle in to about 500~600 kg. According to best estimates, by 2020, China's crude steel production will be about 830 million tons. Subsequent to this point, iron and steel production will be in a stable stage of development. Between the years 2020 and 2030, China's economy will gradually enter the industrial age, and iron and steel production will decline from the peak in steel production per capita. This may be the result of economic optimization, and the industry may witness transfers to another country. Another possibility is that the adjustment of economic structure and the development of alternative materials will lead to declines in consumer demand for steel products. It is expected that by 2030, China's crude steel production will have dropped to 800 million tons.

Alongside the iron and steel industry, the nonferrous metals industry is an energy-intensive and resource-based industry. Since the reforms and opening policy, the nonferrous metal industry has been deeply reformed, adjusted its structure, and achieved some excellent results. This is especially true after the tenth "Five Year Plan" (1994), which demonstrated China's rapid economic development. Since this point, China's nonferrous metal industry has developed rapidly, showing a rapid increase in nonferrous metals demand. Table 4.2 lists primary aluminum production in China, since 1995, as well as primary aluminum production per capita in 2010. China's primary aluminum production was 1,550,000 tons and primary aluminum per capita production reached 11.6 kg.

Figure 4.2 shows the changes in aluminum consumption per capita in the United States. US consumption of aluminum per capita in the early 1970s (1973) had reached a peak level of about 25 kg. At this time, the cumulative consumption of aluminum per capita was up to about 300 kg. 20 years later, US consumption of aluminum per capita was in fluctuation below the basic maintenance level of 22 kg.

Table 4.2 China's Primary Aluminum Production and Primary Aluminum Production Per Capita Since 1995

Year	Primary Aluminum Production (×10,000 tons)	Rate of Growth (%)	Population (×100 million)	Primary Aluminum Production Per Capita (kg)
1995	187		12.11	1.5
1996	190	1.6	12.24	1.6
1997	217	14.2	12.36	1.8
1998	243	12.0	12.48	1.9
1999	281	15.6	12.58	2.2
2000	299	6.4	12.67	2.4
2001	358	19.6	12.76	2.8
2002	451	26.1	12.85	3.5
2003	596	32.2	12.92	4.6
2004	667	11.9	13.00	5.1
2005	781	17.0	13.08	6.0
2006	935	19.8	13.14	7.1
2007	1256	34.3	13.21	9.5
2008	1318	4.9	13.28	9.9
2009	1285	−2.5	13.34	9.6
2010	1550	20.6	13.41	11.6

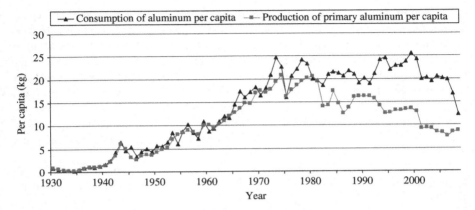

Figure 4.2 Consumption of aluminum per capita and production of primary aluminum per capita in the United States.

Table 4.3 China's Cement Production and Consumption Per Capita from the Year 1990

Year	1990	1995	2000	2005	2009
Cement production (1000 t)	21	47.6	59.7	106.4	163
Annual cement consumption per capita (kg)	183	393	471	806	1221
Cumulative cement consumption per capita (t)	2.26	3.79	5.97	8.69	12.17

For a calculation of the cumulative cement consumption per capita, set the year to more than 100 kg cement consumption per capita for the calculation of the initial year. China reached 100 kg of cement consumption for the first time in the year 1983, achieving a level of 105 kg.
Source: China Cement Industry Association.

The acceleration of industrialization and urbanization will fully stimulate rapid development industries such as real estate and automotive. The expected aluminum consumption per capita in China will peak around 2020 at about 24~25 kg, with about a total of 36,500,000 tons of total consumption. Considering production of recycled aluminum, imports and exports of primary aluminum production in China are estimated to reach about 28,000,000 tons by the year 2020. After the year 2020, economic development will enter the steady development stage, and real estate, automotive, and other downstream industries will show slower growth rates. Also, due to advancements, structural adjustments, and other factors, the aluminum consumption level will continue to be reduced, eventually reaching 25,000,000 million tons.

From the prospective of the materials for construction industry, China is already the world's largest cement producing and consuming country over the past 20 years. Table 4.3 shows China's cement production and consumption per capita from the year 1990.

China is currently undergoing a period of accelerated development in industrialization and urbanization. China is currently in a stage of rapid growth in cement consumption. This is due to a gradually increasing urbanization rate, and the expansion of the total economy. Cement consumption is also expected to continue to rise. By taking into account that the cement standard differs between China and other foreign countries, some experts believe that China's cumulative cement consumption per capita has reached 22~24 tons. This cement consumption will eventually reach a point of saturation, and the consumption rate will gradually slow down. As of 2009, China's cement output was about 1.63 billion tons, cement consumption per capita was 1221 kg, and the cumulative cement consumption per capita was 12.17 tons. Therefore, the future of cement consumption in China has great potential.

According to a correlation analysis over many years, urbanization is positively correlated with demand for cement. If the urbanization rate is increased by 1%, the demand for cement will increase by about 100 million tons. Due to the fact that cement cannot be preserved for extended periods, the production and marketing rate for cement has remained at within the 9~98% range. Thus, from the perspective of the national total, apparent cement consumption and the cement production rate are basically the same. By the best estimates, cement production is expected to reach 2.6 billion tons by 2020, and by 2030, it will be about 3.1 billion tons.

4.4 Regional Economic Layout Optimization

Currently, in the eastern part of China, a majority of the provinces (autonomous regions, municipalities) are considered as development optimization zones. In the central, western, and northeastern regions of China, a majority of the provinces (autonomous regions and municipalities) are the focused areas of development or restricted development zones. The eastern regions, especially the metropolitan areas in the east, are already in the process of transition from the industrialized advanced stage to the post-industrialization era. Here, we have seen both advances in economic structure, and a trend at becoming more service oriented. The central region and the northeast region are still in the latter stage of industrialization, and the industrialization and urbanization process in those regions should continue to further accelerate. The development of the western region is relatively slow and is subject to limitations caused by natural conditions and the fact that the ecological environment is relatively fragile. This is especially true in the northwest arid regions. Priority should be given to protecting the ecological environment and to developing industries with local advantages for industrial development. This holds the keys to success for the future development of the ecological industry. Under the guidance of a national and regional policy, the positioning of the "Four Regional Sections" should be as follows: the eastern region will continue to stay ahead in technology, human resources, capital, openness, and the level of urbanization; the central region should benefit from its geographic position during the program called "Uniting the East with the West"; the western region will promote the development of industries with local advantages as a development strategy; the northeast will be committed to the upgrading of traditional industries and technologies. Specifically, these regional advantages, and their development trends, will have the following characteristics:

1. *The eastern region*: The eastern coastal areas of China are the most economically developed regions. They have the advantages of capital, technology, personnel, policies, and strong socioeconomic conditions. The Yangtze River Delta, Pearl River Delta, Beijing, Tianjin, and other places here still demonstrate China's most dynamic economy. The potential of the eastern region will continue to lead the national economic trend. The development of the Eastern coastal areas should help to improve the capabilities of independent innovation, enhance international competitiveness, and be the first to achieve optimization from the economic structure upgrading. This will be focused on improving the quality and efficiency for the utilization of foreign capital. It will also strive to improve the level and position of the global division of labor and help to enhance the international competitiveness of local industry.

The Yangtze River Delta, Pearl River Delta, and Bohai Sea Region are the core areas of the Eastern Region. In the twelfth Five Year Plan, and even over the long term, the targeted development path will be to enhance the core competitiveness of high-tech industries in these areas. This will eventually help the region to become an important part of the layout of the global high-tech industry. The plan will also help to promote China's industrial sectors by factor-driven to innovation-driven change. For these current development characteristics, the Yangtze River Delta region is expected to take the lead in China's high-tech R&D and advanced manufacturing. It will rely on the industrial base of the Yangtze River Delta region. It will also draw upon its technological strengths and geographic conditions. This region has the advantage of being a well-developed private economy. It can also easily attract high-tech enterprises with innovation capability. With Shanghai as the leader, and Southern Jiangsu, and Northern Zhejiang as the support group, the surrounding cities expansion area will lay a solid foundation for China's strategic development of emerging industries. This development will be aided by the area's innovative and advanced manufacturing system. The Pearl River Delta region has a complete set of industries, manufacturing capacity, and the advantages of well-developed export-oriented economy. It will develop into a global high-tech manufacturing area by consolidating and improving the electronic information product manufacturing industry. It will also vigorously develop capital and the technology-intensive high-tech industry, and promote regional industrial upgrades. The Bohai Sea Region has some talent and research advantages, and it will actively promote the scientific research system. With the promotion of regional scientific and technological innovation optimal allocation of resources, and by building a regional innovation center, the region will improve the business service system and establish a favorable environment for development. It will also seek to actively cultivate new industries.

Intensive production corresponding to the media to the mode shift and the eastern part of the industrial sector, the region of modern services, service industry will also be developed rapidly. The service industry will help to promote the development of high technology and high value-added basis for other industries. Business flow, logistics, capital flow, and information flow in a modern industrial system will promote industrial upgrading in the east. Information, finance, accounting, legal, consulting, marketing services, as well as various intermediary organizations will be expected to aid the overall development of the service industry, as well as improve its overall level.

2. *The central region*: The central region has the advantage of being connected to both the east and the west. It also has advantages in terms of a strong industrial and agricultural foundation, with plentiful elements, resources, and an abundant labor pool. It is currently

the hub area of the national economic sector. The central region will continue to utilize its agricultural advantages, by vigorously developing a more modern agriculture system based on the consolidation of food production. In the industry sector, the central region has formed a number of distinctive high-tech industry gathering areas. These areas include optoelectronic IT in Wuhan, electronic information, biological pharmacy, and new materials developed in Changsha, Zhu zhou, and Xiangtan. The science, technology, and human resources in these cities, as well as their existing industrial base, should accelerate the development of high-tech industries. This will help to make the central region an industrial innovation area. At the same time, the mineral resources should be rationally developed and the product structure should be rationally adjusted. The light agricultural processing industry should also be vigorously developed along with sideline products as raw materials in order to help improve the region's self-sufficiency.

The central region is also rapidly advancing through the stage of industrialization and urbanization because these processes are being actively promoted. The promotion of urbanization and industrialization is an important driving force of economic and social development in the central region. The central region, as an important grain, energy and raw materials base, as well as an equipment manufacturing base, and a transportation hub, has been referred to as being "3 bases, 1 hub." It is currently progressing very steadily. This hub is nurturing the cores of cities by constructing large-scale production and logistics centers. There has also been steadfast promotion of urban and rural development, and a vigorous push towards urbanization. All of this helps to highlight the urban agglomeration as playing a leading role on the regional economy. Another way to help the industrialization process would be to form the so called "two horizontal, two vertical." This is a special economic zone formed by Yangtze River, Longhai, Beijing-Guangzhou, and Beijing-Kowloon. This can be accomplished by actively promoting the development of an urban agglomeration in areas such as the Wuhan City Circle, Changsha, Zhushou, Poyang Lake Ecological Economic Zone, Wanjiang, and Xiangtan regions. In order to improve industrialization, the level of government services needs to be enhanced. This will also help accelerate the construction of a soft environment. To upgrade the industry there must be support for the ability to actively and effectively undertake the industrial transfer. There must be an acceleration of the development of the modern logistics industry, and large-scale production centers and logistics centers must be built. This will also help to promote the central region and the adjacent coastal areas with regional economic integration. Furthermore, it will help the central, eastern and western regions in terms of food, energy, and raw materials and allow for the establishment of long-term and stable cooperative relations.

3. *The western region*: The western region has rich natural and cultural tourism resources, as well as bridges leading to Central Asia and South Asia (including the ASEAN region). These advantages will be supportive of the promotion of western development. Chengyu, Guanzhong-Tianshui, and the Beibu Gulf Economic Zone are western regions of the three major "growth poles." They have the advantages of rich energy resources and industrial development momentum. During the years 2011–2030, the western region will continue to improve its infrastructure, enhance its self-development capacity, and strengthen overall planning and coordination.

From the prospective of industrial development, the "Three rural" issue is still the bottleneck of the western region's economic and social development. This is the reason why it is taking full advantage of the characteristics of its agricultural resources through the adjustment of the agricultural structure. This will help to develop a modern agricultural system and play an important role in increasing the income of farmers. Also, the western region is focused on the development of the more promising areas in the Sichuan Basin, the Hexi Corridor, Central Tibet, and the Xinjiang Yili River Valley in order to improve its agricultural capacity. It hopes to modernize, and to form a large-scale development process for specialty crops, livestock, and agricultural products. These will then be processed for the market.

From the prospective of the adjustment of the economic structure in the Western Region, (in terms of the development of resource industries), in order to accelerate the development of high-tech industries, they must enter a new path of industrialization and provide support for the agriculture and service industries. That is the reason why the western region is focusing on the development of local industries with local advantages, and promoting the development of key economic zones.

The specialization of the western region will fall to energy-intensive industries and the resource processing industry. The National Energy Administration (NEA) has laid out a detailed plan in the twelfth "Five Year Plan" which positions the energy base of the western region. According to the plan, starting from the twelfth "Five Year," China will gradually form five comprehensive energy bases. These will include: Shanxi, the Ordos Basin, the Southwest,Inner Mongolia, and Xinjiang. By the year 2030, it is expected that the five integrated energy bases in the western region will have the supply capacity to account for 85% of the new capacity. This will constitute the basic framework of China's primary energy and pattern. At the same time, the western region's efforts should be made to extend the resource-based industrial chain and to continue to narrow the gap in the industrial division of labor in the Middle Eastern region.

Apart from energy base construction, the west will also be relying on the central cities of Xi'an, Chengdu, Chongqing, and Lanzhou to actively develop the information, biology, aerospace, new materials, new energy, and other high-tech industries. These centers will also be needed for the new environmental protection industry, and will help with advanced technology in order to achieve industrialization while, at the same time, working on saving the environment.

4. *The northeast region*: The northeast region is the home of China's old industrial bases. It is better equipped for manufacturing, and it is rich in natural resources. Over medium- and long-term development, it will promote optimization and will upgrade traditional industrial structures in order to develop high value-added industries and modern services. This will effectively improve the old industrial base with the help of the implementation of policies. It will also focus on improving the capability of independent innovation in order to further upgrade its industrial structure optimization. This will help the region to obtain larger and more efficient manufacturing equipment. It will also see iron and steel, petrochemical, automobile, and other traditional industries support the high-tech industries, and actively cultivate the emerging potential industry. It will also improve the quality of urbanization and guide the Liaoning Coastal Economic Zone, Shenyang Economic Zone, Tai Qi industrial Corridor, and Changjitu Economic Zone to speed up development. Furthermore, it will accelerate the development of a modern service industry to help to actively develop the production of services.

During the years 2011–2030, China's traditional economic layout will experience significant changes. It will mostly rely on the Eastern Region on economic growth. Regional development will be more evenly distributed, and, changes in structures, as well as improved coordination of regional development will be realized. The development of the emerging growth of the Central and Western regions will lead to regional power usage coordination and steady growth.

4.5   Energy Efficiency and Demand Side Management

Energy conservation is one of the most important measures needed in order for countries to achieve green development. During the period of the eleventh "Five Year," the nation's energy intensity fell by 19.1%. During the twelfth "Five Year" period, China's energy-savings emission reduction targets per unit of energy consumption and carbon dioxide components of the GDP. Energy consumption and carbon dioxide emissions were reduced by 16% and 17%.[2] In addition, the complete carbon dioxide emissions per unit of GDP in 2020 compared to 2005 will be decreased by 40~45%. This shows the priority that has been given to energy reduction.

Over the next 10–20 years, China's economy will maintain a sustained and rapid growth. This will mean a continual and increasing demand for energy and electricity. In the increasingly tense situation in terms of energy resources, China is under tremendous pressure to focus on energy savings. This tremendous pressure has also provided a motivating force for the sound development of China's economy and economic restructuring, while shifting China's overall economic development mode. In the future, energy conservation efforts will continue to strengthen their role in economic and social development and become more significant. First, energy conservation will be firmly established. China has been placing great importance on energy saving initiatives. Binding energy-savings emission reduction targets have been included in economic and social development plans, but this will take various measures to promote to fully carry out. Secondly, industrial restructuring has intensified. To change the high input, high consumption, and the high pollution of the extensive mode of growth, the formation of the industrial structure will be characterized by lower input, lower consumption, lower pollution, lower-carbon, and a reduction in the use of iron and steel, nonferrous metals, materials for construction and energy-intensive industries. Furthermore, there will be an elimination of the outdated production capacity and the pace of product replacement will continue to be accelerated. Thirdly, the energy structure adjustment and optimization will need to be emphasized. This will mean that greater efforts will have to be taken in terms of the development of wind, solar, nuclear, and other clean energies in order to reduce the proportions of fossil energy consumption. Fourth, through R&D and upgrades, energy efficiency will vigorously improve. The use of industrial boilers, industrial kilns, and a variety of motors, fans, pumps, irrigation and drainage machinery, compressors, transformers, and other general equipment for energy saving for the overall energy efficiency to reach international advanced level. The development and promotion of new energy-saving technologies will include high infrared heating technology, and microwave high-temperature technology which will be used in the application of membrane technology in gas separation. Also, there will be a promotion of recycled aluminum and efficient electric furnace steel technology, as well as the development of high efficiency pulverized coal combustion technology. Fifth, the market mechanism will be improved, and will effectively promote energy conservation services. In the future, contract energy management, white certificates, carbon trading, and the energy trading market mechanism will be further improved. This will greatly promote energy conservation work, and carry out the formation of the energy services industry.

Due to the efforts to increase energy conservation and the energy efficient services industry in other countries, China will receive an excellent share of this market. The economy will be boosted by energy service companies, energy-savings equipment, and energy-related industries. Overall, the future of China's energy reduction work will be made more effective. The energy consumption of industrial products and overall electricity consumption levels will be effectively reduced, and the consumption levels of various types of electrical equipment will have different degrees of reduction. In 2020, the energy consumption of Chinas major sectors will reach an advanced level. In addition, the upgrading of the industrial structure will reduce electricity intensity. This will help to slow down the growth of electricity demand to a certain extent. As a whole, a number of measures to promote energy savings will still play a significant role in mitigating electricity demand and load growth.

4.6   The Development of Electric Vehicles

Since the beginning of the twenty-first century, China's rapid macro-economic growth has been promoting the rapid development of the automobile industry. For the years 2001–2009, the average annual growth rate of vehicle production was up to 23.5%. This figure is much higher than GDP growth. China's car ownership has also been increasing annually to more than 63 million vehicles in 2009. Along with the sustained and steady development of the national economy, as well as improvements in income and living standards, the rapid development of the automobile industry will continue for more than a decade. The preliminary estimates show that Chinese car ownership will reach 150 million to 200 million by 2020.[3,4]

Vehicles are major energy consumers in China. China's oil import dependency is more than 50%. As car ownership rises, fuel consumption will also rise. This will pose a tremendous danger to China's energy security. Automobile exhaust emissions will also become an important source of urban air pollution. After the financial crisis ends, the development of new energy vehicles will be the new economic growth points and the best choice for many new industries. They will also be an important way to optimize energy structure and energy security. Electric vehicles will be the main direction for the development of new energy vehicles, and they will play an important role in the future development of the automobile industry.

Electric vehicles include pure electric vehicles, hybrid electric vehicles, and fuel cell electric vehicles. The pure electric vehicles are powered by a secondary battery (such as lead-acid batteries, nickel cadmium batteries, nickel metal hydride batteries or lithium-ion batteries). Hybrid electric vehicles use both an internal combustion engine and an electric motor. As the main source of motor power, the internal combustion engine plays only a supporting role. Hybrid electric vehicles are not free from a dependence on oil resources, and therefore, within the development of electric vehicles they represent a transition model. Fuel cell electric vehicles use fuel cells as the power source for an electric motor. Due to their field of use, cost, high-energy intensity, and other problems, they are not suitable for promotion. Overall, the development of a pure electric vehicle is the main direction for future vehicles in China. Throughout the development of electric vehicles, the hybrid vehicles will gradually upgrade and eventually become pure electric vehicles.

After years of experimentation within electric vehicle research and development in China, the environment for the development of electric vehicles is gradually improving. This is creating an excellent foundation for further industrial development. The Ministry of Finance and other ministries have jointly issued a "Notice on the Private Purchase of New Energy Vehicles and Subsidies." The notice was an important step in promoting the development of electric vehicles. Some of the technical standards and access standards of electric vehicles have been put forward, and they will provide important support for the further development of electric vehicles in China.

"The Automobile Industry Restructuring and Revitalization Plan" issued in 2009, proposed, "In 2012, 500,000 cars will be produced, and they will be pure electric, plug-in hybrids, and common-type hybrids, as well as new increases in energy-efficient automobile production capacity. These new energy car sales will account for about 5% of total passenger car sales." The Chinese Society of Automotive Engineers forecasted that by 2015, new energy car ownership will have reached 1 million. At the same time, according to the New Energy Automotive Industry Development Plan, the years 2011–2015 will be important for the promotion of new energy cars. According to comprehensive initial estimates, by 2015, China's electric vehicles will have a capacity to serve 1 million to 1.5 million people. This would account for 1~1.5% of total car ownership.[5] The future prospect for the development of electric vehicles depends on the amount of government policy support to a certain extent. With electric vehicles, as the technology matures, production costs drop. The gradual improvement of the supporting policies, as well as facilities, market acceptance, low running costs, efficiency, and environmental friendliness will be evident. By 2020, driven in part by smart grid technology, electric vehicle technology is expected to have a great breakthrough. By 2030, electric passenger vehicles will be fully developed. The performance of electric vehicles, the cost of production, and other mitigating factors will also be improved. The electric vehicle industry will keep expanding and it will give the economy a positive impact and provide a growth in electricity demand.

Notes

[1]   Joseph AloisSchumpeter. The theory of economic development; 1912.
[2]   Wen Jiabao. The government works report: the fourth meeting of the 11th national people's congress on March 5, 2011. Beijing: People's Publishing House; 2011.
[3]   Li Yizhong. Ministry of Industry and Information Technology.China's auto industry to seize the opportunity to achieve high-level forum on leaps and bounds in 2009.
[4]   Ouyang Ming Gao. CPPCC Standing Committee. People "Members Lecture Hall."
[5]   China Association of Automobile Manufacturers is expecting car ownership in 2015 to be about 100 million to 150 million, 2010.

5 Scenario Analysis of China's Economic Development in the Year 2030

5.1 Scenario Analysis of the National Economic Development

5.1.1 Simulated Results Under Scenario 1

For the various assumptions under Scenario 1, China will continue to maintain a faster rate of economic growth; the projected GDP rate of growth between the years of 2010–2020 is estimated to be 7.9%, while the rate of growth for the years 2020–2030 is estimated to be 5.4% (Table 5.1). The GDP growth rate will present a downward trend for the future, especially between the years of 2020–2030; the reasons for the large scale slowdown being that post-industrialization, the various basic infrastructures of the country are complete, rates of investments have slowed, consumption have reached a much higher level, growth potential has reached its limits, international trade is balanced, and net import growth has slowed down.

From the point of view of the scale of the economy, by 2020, China's total GDP will have increased to 67.2 billion Yuan (in constant 2005 Yuan), the estimated equivalent of USD 8.2 billion (in constant 2005 $). By 2030, it will reach the scale of 113.7 billion Yuan or the equivalent of USD 17.9 billion.

From the point of view of developmental GDP per capita levels, it will be more than USD 5600 by the year 2020 (in constant 2005 $), and estimated to be USD 9349 by 2030. Overall, by 2030, China will have reached the ranks of the upper-middle income countries and poised to reach the upper income countries. (According to the classification standard of World Bank in 2005, the per capita GNI of upper-middle income countries should be in the range of 3466~10725 USD, and the per capita GNI of the upper income countries should be over 10725 USD.)

From a demand point of view (Table 5.2), under Scenario 1, as international trade gradually achieves a balance and the rate of investment falls, the rate of individual consumption is expected to rise to 46.1% in 2020, a rise in 9.2 percentage points compared to 2010. This figure will rise to 52.3% by 2030. The rise in the rate of individual consumption is an important expression of the rise in the living standards of both the rural and urban population. There are three main factors that have hastened the growth of individual consumption—one is the rise in the level of individual consumption, second is the gradual rise in individual wages as a proportion of the gross national income, and third is the rise in the level of the individual's nonwage incomes.

An Exploration into China's Economic Development and Electricity Demand by the Year 2050.
DOI: http://dx.doi.org/10.1016/B978-0-12-420159-0.00005-9

Table 5.1 Scale of the Economy and GDP Per Capita Under Scenario 1

Index	2010	2020	2030	Rate of Growth 2010–2020 (%)	Rate of Growth 2020–2030 (%)
GDP (1000 billion Yuan)	31.4	67.2	113.7	7.9	5.4
GDP per capita (10,000 Yuan)	2.34	4.67	7.74	7.1	5.2
GDP (1000 billion USD)	3.8	8.2	13.9	7.9	5.4
GDP per capita (USD)	2860	5695	9439	7.1	5.2

Calculated at constant price of 2005, the exchange is based on the 2005 exchange rate of 1 USD is to 8.1917 Yuan: the last two columns in the table represents the average rate of growth over each 10-year period.

Table 5.2 2010–2030 Structure of GDP Based on the Expenditure Approach (%)

Index	2010	2020	2030
Personal expenditure	36.9	46.1	52.3
Government expenditure	15.1	15.4	14.9
Total sum of capital formation	42.5	38.2	33.3
Export	32.0	34.8	34.3
Import	26.5	34.5	34.8

In the model, the assumption is made that income from property as a proportion of individual income gradually increases. This is one area in which policy change has been made by the Party and Central Planning Committee in recent years. Of course, the rise on the level of personal consumption also has a more important reason, and that is the increasing aging of the population, the proportion of the population with strong savings will decrease, and the overall rate of savings will also decrease.

As seen from 3-tier industrial framework, the proportion of primary and secondary industry has been decreasing while the proportion of tertiary industry has been rising steadily. This is very similar to the typical developmental pattern of various countries all over the world. From the simulated results, the proportion of primary industry has fallen to about 6.7% in 2020, and by 2030, fallen further to 5.8%. Between the years of 2010 and 2020, the decrease is about 3.4 percentage points, and for 2020–2030, the drop is about 1 percentage point. The proportion of secondary industry is at about 44.5%. In 2030, it will drop further to about 39.7%.

In Scenario 1, the proportion of tertiary industry has risen from 43% in 2010 to 48.7% in 2020—an overall rise of 5.7 percentage points (Table 5.3). By 2030, it may be able to reach a level as high as 54%. But compared to what is experienced by other countries in the world, this proportion is still rather low. When the GDP per capita reaches 10,000 USD, the proportion of tertiary industry in many countries is above 60% on the average.

Based on the economic growth experience of various countries, as a country becomes more developed, its proportion of nonagricultural industries will gradually increase, and there is a sharp rise, especially in the proportion of tertiary industry.

Table 5.3 Structure of GDP of 2010–2030 (%)

Sector	2010	2020	2030
Primary industry	10.2	6.8	5.8
Secondary industry	46.8	44.5	39.7
Tertiary industry	43.0	48.7	54.5

This is a common developmental pattern. The main factor in the push for the increase in tertiary industry is the changing structure of individual consumption, rise in the proportion of services export, as well as an increased demand for the service industry as a necessary input across the board in various sectors. Additionally, the increase in government expenditure will also lead to an increase in the proportion of tertiary industry. The slowdown in export growth also has important impact on the structure of the tertiary industry. As the main export for China comes from manufacturing, and all other factors being constant, a slower rate of export growth is bound to lower the proportion of secondary industry.

First, with the improvements in technical expertise, the rate of use of resources and energy is constantly improving, the rate of energy consumption keeps decreasing, but additionally, there is a greater reliance on external resources and energy. In the long term, therefore, the mining industries as a proportion of GDP will gradually decrease, until between 2020 and 2030, the industry's output as a proportion of GDP will be at 6.2% and 5.4%, respectively (Table 5.4).

Second, within the industrial sector, consumption goods and intermediate goods as a proportion of GDP will be lowered significantly, and the decrease in value of capital goods as a proportion of the GDP will be at a lower rate, while its proportion in relation to consumption goods will be relatively higher. In 2020, the rate of the increase in the value of capital goods and consumption goods as a proportion of GDP will be at 12.4% and 7.9%, respectively. According to Hoffman's measurements, by 2020, when the proportion of the increase in value of China's capital goods is significantly higher than consumption goods, China has basically finished the process of industrialization!

5.1.2 Simulated Results Under Scenario 2

The level of economic development in Scenario 2 is much higher than in Scenario 1, as given in Table 5.5. In Scenario 2, China can still continue to maintain a higher rate of growth, the rate of growth of GDP between 2010 and 2020 is about 8.1%, higher than in Scenario 2 by 0.2 percentage points. Rate of GDP growth is at about 5.6% during 2020–2030, which is also higher than in Scenario 1 by 0.2 percentage points. The rate of change of GDP is similar to Scenario 1, and is also because of the slowdown in investment, consumption, and net export post-industrialization. The overall rate of GDP growth presents a decreasing trend.

In Scenario 2, in 2020, China's GDP will be a total of 68.4 billion Yuan (about 8.4 billion USD); by 2030, it will have reached 118.0 billion Yuan (about 14.4 billion

Table 5.4 Detailed Changes in the Structure of GDP During 2010–2030 (Value Added as a Proportion of GDP, %)

Sector	2010	2020	2030
Primary industry	10.2	6.8	5.8
Agriculture	10.2	6.8	5.8
Secondary industry	46.8	44.5	39.7
Mining	6.0	6.2[a]	5.4
Coal	2.0	2.5	2.7
Oil and natural gas	2.7	2.6	1.8
Selected metals extraction	0.8	0.7	0.6
Selected nonmetals extraction	0.4	0.4	0.3
Manufacturing	34.8	32.0	28.4
Consumption Goods	**8.2**	**7.9**	**7.6**
Foods	2.9	2.5	2.4
Textiles	1.6	1.6	1.5
Fashion	1.5	1.6	1.7
Timber processing, furniture	0.8	0.8	0.7
Paper and publishing	1.5	1.4	1.4
Intermediate Goods	**14.5**	**11.7**	**9.1**
Oil processing and refining	1.1	0.7	0.4
Chemicals	3.1	2.2	1.5
Nonmetallic products	3.6	3.5	3.0
Metals refining	3.6	2.9	2.2
Electricity	2.8	2.1	1.6
Gas	0.1	0.1	0.1
Water	0.2	0.2	0.2
Capital Goods	**12.0**	**12.4**	**11.7**
Metallic goods	1.3	1.3	1.2
Mechanical industry	3.3	2.9	2.5
Transport facilities	2.0	2.1	2.0
Electrical and mechanical	1.7	1.8	1.7
Electronic communications equipment	2.1	2.8	2.8
Devices and measurement devices	0.4	0.5	0.5
Other manufacturing	1.2	1.0	0.9
Construction	**6.1**	**6.3**	**5.9**
Tertiary industry	43.0	48.7	54.5
Transport	6.1	6.1	6.6
Postal and telecommunications	2.7	2.7	2.7
Commerce	6.9	6.9	7.4
Food and beverage	2.3	2.5	2.9
Finance and insurance	3.9	4.8	6.0
Property	4.9	5.1	5.0
Social services	5.6	7.3	9.1
Educational and medical	6.3	9.0	11.0
Government	4.2	4.2	3.8
Total	100.0	100.0	100.0

[a]The German economist, Hoffman summed up the changes within the three industrial sectors, and discovered that any country undergoing industrialization must necessarily experience four stages. This is known in the field of economics as the four stages of Hoffman's theorem of experiential industrialization: (1) Among manufactury goods, the consumption sector will be central; (2) the growth rate of capital goods is faster than that of consumer goods and reaches a point of 50% of the net worth of the consumer goods industry; (3) The capital goods industry continue to grow rapidly, almost reaching equilibrium with the consumer goods industry; (4) The capital goods industry takes over leading position. We can calculate the ratio of consumption goods and capital goods to determine the stage of industrialization that this country (area) is in. This ratio is also known as Hoffman's ratio.

Table 5.5 Scale of the Economy and GDP Per Capita Under Scenario 2

Index	2010	2020	2030	Rate of Growth 2010–2020 (%)	Rate of Growth 2020–2030 (%)
GDP (1000 billion Yuan)	31.4	68.4	118.0	8.1	5.6
GDP per capita (10,000 Yuan)	2.34	4.75	8.03	7.3	5.4
GDP (1000 billion USD)	3.8	8.4	14.4	8.1	5.6
GDP per capita (USD)	2860	5802	9799	7.3	5.4

Calculated at constant price of 2005; the exchange is based on the 2005 exchange rate of 1 USD is to 8.1917 Yuan: the last two columns in the table represent the average rate of growth over each 10-year period.

Table 5.6 Structure of GDP Based on the Expenditure Approach During 2010–2030 (%)

Index	2010	2020	2030
Personal expenditure	36.9	46.8	53.2
Government expenditure	15.1	14.9	14.3
Total sum of capital formation	42.5	37.5	32.1
Export	32.0	34.8	34.8
Import	26.5	34.0	34.4

USD), higher than in Scenario 1 by 1.2 billion Yuan and 4.3 billion Yuan, respectively. In 2020, the GDP per capita will reach about 48,000 Yuan (about 5802 USD) and in 2030 will hit 80,000 Yuan (about 9799 USD), higher than in Scenario 1 by 1.9% and 3.8%, respectively.

In Scenario 2, the structure of investment, consumption, as well as import and export is more balanced; the proportion of personal consumption has risen compared to Scenario 1 (Table 5.6). By 2020, the proportion of personal consumption will rise to 46.8%, and by 2030, will further rise to 53.2%, higher than in Scenario 1 by about 0.7 and 0.9 percentage points, respectively. The higher proportion of personal income compared to Scenario 1 is reflective of a better balance between the proportions of internal investments and consumption. The people get better value from the economic growth. One very important feature of the changing developmental mode is that instead of being dependent on investment and export, the economy is led by the three-horse carriage of consumption, investment, and export simultaneously. The stimulating effect of consumption is especially important.

In Scenario 3, The proportion of tertiary industry is higher than in Scenario 1, for example, in Table 5.7, by 2020, the proportion of tertiary industry would have hit 50.4%, 1.7 percentage points higher than in Scenario 1; and by 2030, the proportion of tertiary industry would have reached 57.4%, 2.9 percentage points higher than in Scenario 1. As can be seen, the rate of growth of the tertiary industry is much faster than in Scenario 1. This is also reflective of an important feature of a changing developmental mode, from a previous method of relying heavily on manufacturing

Table 5.7 Structure of GDP of 2010–2030 (%)

Sector	2010	2020	2030
Primary industry	10.2	6.9	5.9
Secondary industry	46.8	42.7	36.7
Tertiary industry	43.0	50.4	57.4

for quick economic growth, to an upgraded industrial structure that is optimized, based on the simultaneous developments of both the manufacturing and services sectors.

The changes that can be seen in the detailed industrial structure (Table 5.8) are quite similar to that of Scenario 1. First, the mining industry as a proportion of GDP is decreasing steadily, and compared to Scenario 1, the rate of decrease is greater. For example by 2030, the mining Industry as a proportion of GDP would be 5.4%. Second, within the industrial sector, the increase in the value of consumption goods and intermediate goods as a proportion of the GDP is significantly lower, and the rate of decrease of capital goods is much smaller; its proportion of the GDP relative to consumption goods is also higher than in Scenario 1.

5.1.3   Simulated Results Under Scenario 3

In Scenario 3, the rate of economic growth is lower than that of Scenario 1, its rate of growth of GDP being 7.5% and 4.9 %, respectively for the periods of 2010–2020 and 2020–2030. The rate of growth of the GDP is also a downward trend (Table 5.9).

In Scenario 3, by 2020, GDP would have reached 64.7 billion Yuan (about 7.9 billion USD). By 2030, it would have reached 104.4 billion Yuan (about 12.7 billion USD), lower than in Scenario 1 by 2.3 Yuan and 9.3 Yuan, respectively. GDP per capita at 2020 and 2030 are at 4.5 billion Yuan and 7.1 billion Yuan and lower than in Scenario 1 by 3.7% and 8.2%, respectively.

Calculated at constant price of 2005, the exchange is based on the 2005 exchange rate of 1 USD is to 8.1917 Yuan: the last two columns in the table represent the average rate of growth over each 10-year period.

Because of greater external demand as a proportion of China's GDP, the deceleration in the rate of growth of all exports is one of the important influential factors in Scenario 3. Therefore, based on the structure of expenditure approach in Scenario 3, there is a decrease in the proportion of exports, lower than in Scenario 1 by 1 percentage point in 2020 and lower by 1.5 percentage points in 2030 (Table 5.10).

In Scenario 3, because of the slowdown due to adjustments in the economic structure, the proportion of the secondary industry is higher than in Scenario 1; the primary and tertiary industry are correspondingly lower than in Scenario 1. For example, as given in Table 5.11, up to 2020, the proportion of three industries is 6.7%, 44.8%, and 48.5%, respectively. By 2030, it is 5.6%, 40.0%, and 54.4%, respectively.

Table 5.8 Detailed Changes in the Structure of GDP During 2010–2030 (Value Added as a Proportion of GDP, %)

Sector	2010	2020	2030
Primary industry	10.2	6.9	5.9
Agriculture	10.2	6.9	5.9
Secondary industry	46.8	42.7	36.7
Mining	6.0	6.3	5.4
Coal	2.0	2.7	2.9
Oil and natural gas	2.7	2.6	1.8
Selected metals extraction	0.8	0.6	0.5
Selected nonmetals extraction	0.4	0.4	0.3
Manufacturing	34.8	30.2	25.5
Consumption Goods	**8.2**	**7.1**	**6.4**
Foods	2.9	2.6	2.6
Textiles	1.6	1.2	0.9
Fashion	1.5	1.3	1.3
Timber processing, furniture	0.8	0.7	0.5
Paper and publishing	1.5	1.3	1.1
Intermediate Goods	**14.5**	**10.8**	**7.9**
Oil processing and refining	1.1	0.6	0.4
Chemicals	3.1	2.0	1.2
Nonmetallic products	3.6	3.2	2.6
Metals refining	3.6	2.7	1.9
Electricity	2.8	1.9	1.4
Gas	0.1	0.1	0.1
Water	0.2	0.2	0.2
Capital Goods	**12.0**	**12.3**	**11.3**
Metallic goods	1.3	1.1	0.9
Mechanical industry	3.3	2.7	2.1
Transport facilities	2.0	2.0	1.9
Electrical and mechanical	1.7	1.9	1.8
Electronic communications equipment	2.1	3.1	3.2
Devices and measurement devices	0.4	0.6	0.6
Other manufacturing	1.2	0.9	0.8
Construction	**6.1**	**6.2**	**5.7**
Tertiary industry	43.0	50.5	57.4
Transport	6.1	6.6	7.4
Postal and telecommunications	2.7	2.8	2.9
Commerce	6.9	7.6	8.5
Food and beverage	2.3	2.6	3.1
Finance and insurance	3.9	5.0	6.2
Property	4.9	5.2	5.1
Social services	5.6	7.6	9.7
Educational and medical	6.3	8.9	10.9
Government	4.2	4.1	3.6
Total	100.0	100.0	100.0

Table 5.9 Scale of the Economy and GDP Per Capita Under Scenario 3

Index	2010	2020	2030	Rate of Growth 2010–2020 (%)	Rate of Growth 2020–2030 (%)
GDP (1000 billion Yuan)	31.4	64.7	104.4	7.5	4.9
GDP per capita (10,000 Yuan)	2.34	4.49	7.10	6.7	4.7
GDP (1000 billion USD)	3.8	7.9	12.7	7.5	4.9
GDP per capita (USD)	2860	5487	8672	6.7	4.7

Calculated at constant price of 2005, the exchange is based on the 2005 exchange rate of 1 USD is to 8.1917 yuan: the last two columns in the table represents the average rate of growth over each 10-year period.

Table 5.10 Structure of GDP Based on the Expenditure Approach During 2010–2030 (%)

Index	2010	2020	2030
Personal expenditure	36.9	46.0	52.0
Government expenditure	15.1	15.4	14.9
Total sum of capital formation	42.5	38.1	33.3
Export	32.0	33.8	32.8
Import	26.4	33.3	33.2

Table 5.11 Structure of Industrial Sectors During 2010–2030 (%)

Sector	2010	2020	2030
Primary industry	10.2	6.7	5.6
Secondary industry	46.8	44.8	40.0
Tertiary industry	43.0	48.5	54.4

From within the industrial sector, (Table 5.12), mining as a proportion of the GDP first rose before falling, and from 6.2% in 2010, it is raised to 6.4% in 2015. By 2030, it will have fallen to 5.7%. Within the mining industry, coal has a greater proportion and it continues to rise, but oil and natural gas is correspondingly lower. This is in line with the state of unevenness of natural resources in China.

5.2 Analysis of Scenarios of Regional Economic Development

In line with the results of the economic development within the country, the economic developments of the various regions remain within the three scenarios that have been outlined. Only the simulated results of the economic development of the various regions under Scenario 2 will be dealt with here.

Table 5.12 Detailed Changes in the Structure of GDP During 2010–2030
(Value Added as a Proportion of GDP, %)

Sector	2010	2020	2030
Primary industry	10.2	6.7	5.6
Agriculture	10.2	6.7	5.6
Secondary industry	46.8	44.8	40.0
Mining	6.0	6.4	5.7
Coal	2.0	2.6	2.8
Oil and natural gas	2.7	2.7	2.0
Selected metals extraction	0.8	0.7	0.6
Selected nonmetals extraction	0.4	0.4	0.3
Manufacturing	34.8	32.1	28.4
Consumption Goods	**8.2**	**8.0**	**7.7**
Foods	2.9	2.5	2.4
Textiles	1.6	1.6	1.5
Fashion	1.5	1.6	1.7
Timber processing, furniture	0.8	0.8	0.7
Paper and publishing	1.5	1.5	1.4
Intermediate Goods	**14.5**	**11.9**	**9.4**
Oil processing and refining	1.1	0.7	0.5
Chemicals	3.1	2.2	1.6
Nonmetallic products	3.6	3.5	3.1
Metals refining	3.6	3.0	2.3
Electricity	2.8	2.2	1.6
Gas	0.1	0.1	0.1
Water	0.2	0.2	0.2
Capital Goods	**12.0**	**12.2**	**11.4**
Metallic goods	1.3	1.3	1.2
Mechanical industry	3.3	2.9	2.5
Transport facilities	2.0	2.1	2.0
Electrical and mechanical	1.7	1.8	1.7
Electronic communications equipment	2.1	2.7	2.6
Devices and measurement devices	0.4	0.5	0.5
Other manufacturing	1.2	1.0	0.9
Construction	**6.1**	**6.3**	**5.9**
Tertiary industry	43.0	48.5	54.4
Transport	6.1	6.1	6.6
Postal and Telecommunications	2.7	2.6	2.7
Commerce	6.9	6.9	7.4
Food and beverage	2.3	2.5	2.9
Finance and insurance	3.9	4.8	6.0
Property	4.9	5.1	5.0
Social services	5.6	7.3	9.1
Educational and medical	6.3	9.0	10.9
Government	4.2	4.2	3.8
Total	100.0	100.0	100.0

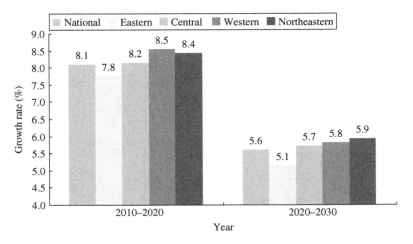

Figure 5.1 GDP growth rate of various regions at various stages for 2010–2030.

5.2.1 Comparison Across Regions

Figure 5.1 shows the rate of GDP growth of the Eastern, Central, Western, and Northeastern regions. From the figure, the various regions of China will still maintain a faster rate of economic growth in 2020. After 2020, the economic growth of the various regions is significantly slower. At the regional level, disparity in economic growth across different regions is still quite apparent. The economic growth of the Central and Western regions is much faster than in the east. The economic growth rate of the less developed regions is faster than that of the developed regions. The performance of the regional economic development appears to be more subdued.

There are three main reasons for the faster growth of the economic development in the Central and West regions as compared to the East. First, the energy resources of the Central and West regions are in greater abundance and hence, the capital investment for energy-related industries is lower and its development is faster. As China's economy continues to grow, there is increasing demand on the environment and resources, and the Central and Western regions are the main depositories of China's resources. Therefore, as the issue of the ever increasing constraints of energy supply on the economy grows in its importance, the lower investment costs of the central and west regions in this aspect allows them to enjoy a certain comparative advantage. Second, the labor productivity of the Central and West regions have been proceeding at a faster pace. Ever since the reforms and the opening up of the country, the coastal regions in the East have been developing faster, mainly due to its reaping the benefits of the country's reforms and open door policies. But the labor force in the Central and West regions have been moving toward the east in large numbers, providing a constant source of low cost labor. This is also an important factor. As the surplus labor force in the rural areas gradually decreases, but the large labor supply that has been shifting to the east is not allowed to settle there and be localized, much of the labor force in the central and west regions have been more inclined to find jobs within

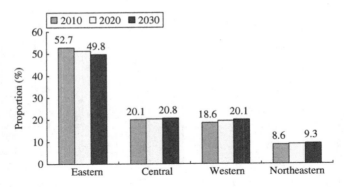

Figure 5.2 Proportion of the country's GDP of the various regions in the middle and long term.

the region itself. Therefore, by comparison, the growth of the labor force in the central and western regions has been faster. Third, the level of management and technical expertise has been relatively lower in the central and western regions. Through learning from the East region as well as overseas, they have been able to raise their level of productivity at a much faster rate. Also, the Central and West regions enjoy a certain advantage of hindsight. This is one of the reasons for the faster growth in the Central and West regions.

From what can be seen, of the proportion of the country's GDP contributed by the various regions, (as shown in Figure 5.2), the proportion of the nation's GDP from the East region is significantly higher than that of the other three; the central and west regions are evenly matched, while the Northeast region forms the lowest proportion of the nation's GDP. From what can be seen in the changes in the proportion of the GDP, the central and west regions display an upward trend because of their faster economic growth. It's the same for the Northeast region. The East region is the only region that displays a falling trend in its proportion of the national GDP.

By 2020, the percentages of these four regions as a proportion of the national GDP—East, Central, Western, and Northeast are 51.3%, 20.3%, 19.5%, and 8.9%, respectively. Between 2010 and 2020, the percentage proportion of the national GDP of the East region will have fallen by 1.4 percentage points. The Central, West, and Northeast regions will all show a rise of 0.2, 0.9, and 0.3 percentage points, respectively. Between the years of 2020–2030, the percentage of the East region's proportion of the national GDP will fall by 1.5 percentage points, and the central, west and Northeast Regions will rise by 0.5, 0.6, and 0.4 percentage points, respectively. By 2030, the figures for the contributions of the four major regions—East, Central, West and Northeast to national GDP will be 49.8%, 20.8%, 20.1%, and 9.3%, respectively.

From the structure of the industrial sectors (Figure 5.3), all the four regions show a decreasing trend for the proportions of their primary and secondary industries and a constant rise in the proportion of their tertiary industry from the period of 2010–2030. In the same period, the proportion of the primary industry for the East, Central, West and Northeast regions all show a drop of 2.8, 6.5, 5.3, and 2.0 percentage points, respectively; the proportion of secondary industry all fall by 10.3, 4.7,

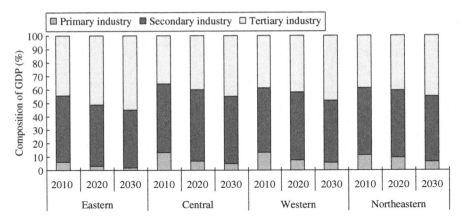

Figure 5.3 Industrial sector changes for the four regions at various stages for 2010–2030.

6.4, and 10.9 percentage points, respectively; and their tertiary industry rise by 13.1, 11.2, 11.7, and 12.9 percentage points, respectively.

The proportion of the secondary industry of the various regions continues to fall, and this has to do with the overall gradual completion of China's progress in the industrialization process. Overall as a country, we are already in the middle of the later stages of industrialization and moving toward the later stages in the middle to long run. One characteristic of the later stage industrialization is the decline of the proportion of the industrial sector. Even though there is great disparity in terms of the level of industrialization across the country, many of the regions are still in the initial or middle stages of industrialization. But because of the mobility of goods within the country, the rising trends of the industrial sectors of the various regions are not the same as if they were independent, autonomous industrial sectors.

The East Region

Due to the fact that the East region has already entered the later stages on industrialization, its economic growth will be significantly slowed down for the next 10–20 years, and among the four regions, the rate of growth of GDP for the East region is the lowest. (For the simulated results of the economic growth of the region, see Table 5.13.) For 2010–2030, it is projected that the annual growth rate of the East region's GDP will be at 7.8%, lower than the national average by 0.3 percentage points. For 2020–2030, the average GDP growth rate of the East region is at 5.1%, lower than the national average by 0.5 percentage points.

From a provincial level, Beijing and Shanghai have already completed their industrialization. They will focus on the high tech and service industries, and their economic growth will be slowed significantly. Compared to the two direct administrative cities of Beijing and Shanghai, the development of Tianjin is slower. But with the incorporation of the coastal area into the nation's developmental strategy, there is greater potential for Tianjin's economic growth in the future. The broad and

Table 5.13 Simulated GDP Results of the Provinces (Cities) in the East Region (at Constant Price of 2005, per 100 million Yuan)

Provinces (Cities) in the East Region	2010	2020	2030	2010–2020 Growth Rate (%)	2020–2030 Growth Rate (%)
Beijing	10,019	21,435	34,314	7.9	4.8
Tianjin	5666	14,102	25,337	9.5	6.0
Hebei	15,278	29,821	45,993	6.9	4.4
Shandong	29,255	63,027	10,7148	8.0	5.4
Shanghai	11,990	23,773	37,576	7.1	4.7
Jiangsu	27,877	61,766	104,635	8.3	5.4
Zhejiang	19,562	37,863	59,535	6.8	4.6
Fujian	9785	22,703	38,513	8.8	5.4
Guangdong	30,398	62,793	103,544	7.5	5.1
Hainan	1350	2813	4617	7.6	5.1
East region	161,180	340,096	561,212	7.8	5.1

GDP is calculated according to 2005 prices.
The last two columns of the table show the average GDP growth rate of the various regions for 2010–2020 and 2020–2030.

extensive manner of economic development in Hebei will become increasingly disparate with greater economic restrictions in the days to come. The basic and manufacturing industries of Shan Dong enjoy a distinct advantage, and its economy will continue to see steady growth. Zhejiang will hasten the development of its service industry, especially in terms of the modern services. Economic growth will slow down. Fujian enjoys the benefits of the developments of the "West Coast Economic Zone" and will become an important gateway for servicing the middle and western regions; as well as an important base for advanced manufacturing. Its economic growth will be faster. Guangdong, being the nation's greatest process trading province, with the country's rising cost of labor, will see significantly slower economic growth. The main industries for Hainan are its tourism and the related modern services. Its economic growth will also be slower. Tianjin's GDP growth rate will be 9.5%, while Beijing, Hebei, Shandong, Shanghai, Zhejiang, Guangdong, and Hainan will all experience a GDP that is lower than the national average, the rate of growth will be between 6.8% and 8%. Among these, Beijing's rate will be 0.2 percentage points lower than the national average, while for Shanghai, it will be lower by 1 percentage point. During 2020–2030, only Tianjin will enjoy an economic growth that is higher than the national average. Its growth rate will be 6.0%. The other provinces (cities) will be lower than the national average and be between 4.4% and 5.4%.

From the GDP changes of the various provinces of the East region (as shown in Figure 5.4), for the next 20 years, the main sources of economic growth in the East region will come from the four provinces of Shandong, Jiangsu, Guangdong, Zhejiang. 2010–2020, the GDP contributions of the four provinces are a total of 66.1% (Shandong 18.9%, Jiangsu 18.9%, Guangdong 18.1%, Zhejiang 10.2%). For 2020–2030, the GDP contributions of the four provinces will be nearly 68.0% (Shandong 20.0%, Jiangsu 19.4%, Guangdong 18.4%, Zhejiang 9.8%).

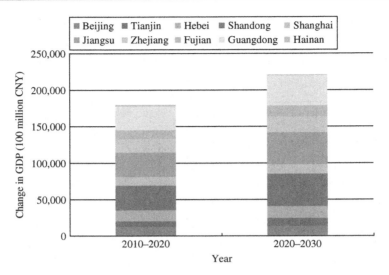

Figure 5.4 GDP changes in the various provinces (cities) of the East region.

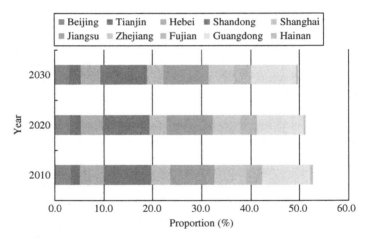

Figure 5.5 Proportion of the national GDP as contributed by the various provinces (cities) in the East region.

As indicated by Figure 5.5, by 2020, the percentage of the East region as a proportion of the national GDP will be at 51.3%; for 2010–2020, this figure will fall by 1.4 percentage points; for 2020–2030, this figure falls another 1.5 percentage points. By 2030, the percentage of the East region as a proportion of the national GDP will be at 49.8%. From the conditions of the various provinces (cities), for 2010–2020, Tianjin, Jiangsu, and Fujian will see a rise in its proportion of the national GDP. The rise will be 0.28, 0.20, 0.23 percentage points, respectively. The other seven provinces

Table 5.14 Structural Changes for the East Region (%)

Province (City)	Industry	2010	2020	2030
Beijing	Primary	0.8	0.3	0.2
	Secondary	21.9	15.5	12.8
	Tertiary	77.3	84.2	87.0
Tianjin	Primary	1.6	0.5	0.3
	Secondary	54.8	49.3	40.8
	Tertiary	43.6	50.2	58.9
Hebei	Primary	11.3	9.6	9.4
	Secondary	53.0	48.3	44.1
	Tertiary	35.7	42.1	46.5
Shandong	Primary	8.5	5.7	5.3
	Secondary	53.6	43.7	33.7
	Tertiary	37.9	50.6	61.0
Shanghai	Primary	0.7	0.5	0.5
	Secondary	39.7	30.8	23.9
	Tertiary	59.6	68.7	75.6
Jiangsu	Primary	5.7	2.2	1.9
	Secondary	51.6	45.5	38.9
	Tertiary	42.7	52.3	59.2
Zhejiang	Primary	4.0	1.2	0.7
	Secondary	49.7	40.6	33.7
	Tertiary	46.3	58.2	65.6
Fujian	Primary	9.2	4.4	3.6
	Secondary	46.2	41.5	34.6
	Tertiary	44.6	54.1	61.8
Guangdong	Primary	4.9	2.6	2.1
	Secondary	46.6	43.2	39.0
	Tertiary	48.5	54.2	58.9
Hainan	Primary	26.2	15.9	13.0
	Secondary	28.9	27.2	23.1
	Tertiary	44.9	56.9	63.9

(cities) will all see a drop in its proportion of the national GDP. Among them, Hebei, Zhejiang, and Guangdong will see the biggest drops—0.50, 0.68, and 0.46 percentage points, respectively. For 2020–2030, Tianjin's proportion of the national GDP will increase while all the other nine provinces (cities) will present a downward trend. The decrease will still be most significant for Hebei and Guangdong.

Table 5.14 lists out the changes in the 3-tier industrial sector structures for the various provinces (cities) in the East region. With the exception of Hainan, there is an all-round drop in the proportion of primary industry, and a rise in the tertiary industry. By 2020, except for Hebei Province, all the other provinces (cities) will see the proportion of its tertiary industry pass the 50% mark. Hubei's proportion of tertiary industry is at 42.1%. By 2030, this figure will be 46.5%, and all the other provinces (cities) will have proportion of tertiary industry that is above 60%. Among them, the percentage of tertiary industries for Beijing and Shanghai are at 87% and 76%, respectively.

Table 5.15 Simulated GDP Results of Provinces in Central Region (at Constant Price of 2005, per 100 million Yuan)

Provinces in the East Region	2010	2020	2030	2010–2020 Growth Rate (%)	2020–2030 Growth Rate (%)
Shanxi	7157	15,317	26,594	7.9	5.7
Anhui	8614	18,661	31,701	8.0	5.4
Henan	18,078	38,769	67,479	7.9	5.7
Hubei	10,703	24,246	43,139	8.5	5.9
Hunan	10,797	24,405	42,982	8.5	5.8
Jiangxi	6152	13,242	22,655	8.0	5.5
Central region	61,501	134,640	234,550	8.2	5.7

GDP is calculated according to 2005 prices. The last two columns of the table show the average GDP growth rate of the various regions for 2010–2020 and 2020–2030.

Central Region

The Central region is mainly a beneficiary of the country's strategy of "raising up the Central" where its economy will get a faster push. The inherent contradictions in "raising up the Central" will be improved. (For the simulated results of the economic growth of the Central region, see Table 5.15.) It is projected that for the years 2010–2020 and 2020–2030, the average GDP growth rate of the Central region will be 8.2% and 5.7%, respectively, both higher than the national average by 0.1 percentage points.

On the provincial level, Shanxi is supported by resource-intensive industries. Its economic growth will drop significantly. The shifts in industry that are taking place in Anhui, Henan, Hubei, Hunan, and Jiangxi will play an important role. For 2010–2020, the GDP growth rate of Hubei and Hunan will be higher than the national average, at about 8.5%, whereas for Shanxi, Anhui, Henan, and Jiangxi, their growth will be lower than the national average, all at around 8.0%. For 2020–2030, the growth rate of the provinces in the Central region will be close to the national average which is between 5.4% and 5.9%.

From the GDP changes of the various provinces of the Central region (as shown in Figure 5.6), for the next 20 years, the main sources of economic growth in the Central region will come from the three provinces of Henan, Hubei, Hunan. 2010–2020, the GDP contributions of the three provinces are a total of 65.3% (Henan 28.3%, Hubei 18.5%, Hunan 18.5%). For 2020–2030, the GDP contributions of the three provinces will be nearly 66.2%. As shown in Figure 5.7, by 2020, the proportion of the national GDP contributed by the Central region will be 20.3%; during 2010–2020, this figure will have risen by 0.2 percentage points, and by 0.5 percentage points between 2020 and 2030. By 2030, the percentage of the Central region as a proportion of the national GDP will be 20.8%.

Table 5.16 lists the changes in the economic structure. For the Central region, with the exception of Shanxi, the proportion of primary industry for the other provinces is much higher. They are all currently at above 11%. The highest is Hunan, in which primary industry is as high as 16.2%. For the period of simulation, the proportion of primary industry of Hubei and Hunan decreased, but the

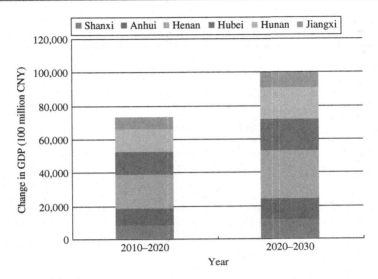

Figure 5.6 GDP changes in the various provinces of the Central region.

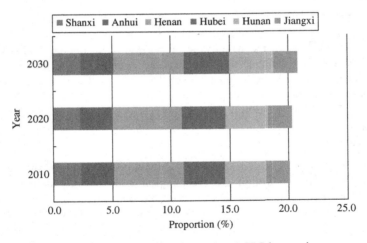

Figure 5.7 The Central region's proportion of the national GDP by provinces.

rate of decrease is quite small. By 2030, the proportion of primary industries of these two areas will be at 9.1% and 10.9%, respectively. Significantly higher than the national average of 5.9%, the dip for the proportion of primary industry for Anhui and Hunan is quite great. From 2010 to 2030, they fell by 9.3 and 8.2 percentage points, respectively. As for tertiary industry, by 2030, the four provinces of Anhui, Hubei, Hunan, and Jiangxi would have tertiary industry proportions of more than 50%. And for Henan, it may be as high as 49%. The proportion of Shanxi's secondary industry has always been quite high. This has to do with its rich coal resources giving it a comparative advantage for other related energy

Table 5.16 Changes in the Industrial Sectors of the Central Region (%)

Province	Industry	2010	2020	2030
Shanxi	Primary	3.1	1.3	0.9
	Secondary	61.9	58.1	53.1
	Tertiary	35.0	40.6	46.0
Anhui	Primary	12.7	5.5	3.4
	Secondary	46.8	43.2	38.2
	Tertiary	40.5	51.3	58.4
Henan	Primary	11.5	5.0	3.3
	Secondary	57.4	54.2	47.7
	Tertiary	31.1	40.8	49.0
Hubei	Primary	14.2	10.7	9.1
	Secondary	41.6	35.2	28.7
	Tertiary	44.2	54.1	62.2
Hunan	Primary	16.2	12.5	10.9
	Secondary	43.1	38.3	32.7
	Tertiary	40.7	49.2	56.4
Jiangxi	Primary	13.9	8.3	6.9
	Secondary	51.1	47.3	41.0
	Tertiary	35.0	44.4	52.1

and heavy industries. On the whole, the proportion of secondary industry for the Central region will be higher than the national average in the middle and long term. This has to do with the hastened development of the Central region as well as the continual relocations of certain industrial sectors, especially the industrial sector from the East to the Central region.

West Region

Under the push of the "develop the West" economic strategy, for the next 10–20 years, the economic growth of the West region will be faster than the national average. (For simulated results of the economic growth of the West region, see Table 5.17). For 2010–2020, the rate of GDP growth for the West region is at 8.5%, which is higher than the national average by 0.4 percentage points for the same period. For 2020–2030, the average rate of GDP growth for the West region is at 5.8%, which is higher than the national average by 0.2 percentage points.

On a provincial (area, city) level, the economy of Inner Mongolia is mainly sustained by "resource-type" industries. Affected by the macrolevel adjustments by the country, its economy, while growing at a fast rate, will see periodic downward adjustments. Sichuan and Chongqing, because of the push provided by the "develop the West" strategy as well as the establishment of economic areas, will have quite fast economic growth. Shaanxi province enjoys a special geographical advantage in the country's strategic thrust. Its economy grows at a fast yet steady rate. Gansu, Qinghai, and Ningxia because of the constraints brought about by its high energy intensity industrial developments will have a lower rate of economic growth. Xinjiang, Tibet,

Table 5.17 Simulated GDP Results of the Provinces (Cities) in the West Region (at Constant Price of 2005, per 100 million Yuan)

Provinces (Cities) in the West Region	2010	2020	2030	2010–2020 Growth Rate (%)	2020–2030 Growth Rate (%)
Inner Mongolia	8539	21,834	38,910	9.8	5.9
Sichuan	11,697	26,400	46,810	8.5	5.9
Chongqing	4937	11,834	20,739	9.1	5.8
Shaanxi	6723	16,239	30,102	9.2	6.4
Gansu	3030	5979	9697	7.0	5.0
Qinghai	924	1891	3139	7.4	5.2
Ningxia	1073	2215	3705	7.5	5.3
Xinjiang	3895	8468	15,993	8.1	6.6
Tibet	448	1022	1823	8.6	6.0
Guangxi	6936	15,627	27,530	8.5	5.8
Guizhou	3274	7016	11,817	7.9	5.4
Yunnan	5311	10,289	16,355	6.8	4.7
West region	56,787	128,814	226,620	8.5	5.8

GDP is calculated according to 2005 prices.
The last two columns of the table show the average GDP growth rate of the various regions for 2010–2020 and 2020–2030.

and Guizhou because of the government's push to develop energy industries will also see faster economic growth. For 2010–2020, the disparity between the various provinces (area, city) in the West region will be quite wide. Among them, Inner Mongolia, Sichuan, Chongqing, Shaanxi, Xinjiang, Tibet, and Guangxi will all see GDP growth at above national average, within the range of 8.1–9.8%, while Gansu, Qinghai, Ningxia, Guizhou, and Yunnan will see economic growth rates that are below the national average between 6.8% and 7.9%. For 2020–2030, there will not be many changes to the mode of economic growth of the provinces (areas, cities) in the West region. Inner Mongolia, Sichuan, Chongqing, Shaanxi, Xinjiang, Tibet, and Guangxi will all see GDP growth at above national average, within the range of 5.8–6.6%, while Gansu, Qinghai, Ningxia, Guizhou, and Yunnan will see economic growth rates that are below the national average between 4.7% and 5.4%.

From the GDP changes of the individual provinces in the West region (area, city), as shown in Figure 5.8, for the next 20 years, the main contributions to GDP growth in the West region will be mainly from Inner Mongolia, Sichuan, Shanxi, and Guangxi. For the next 20 years, the percentage total contributions of these provinces (areas, cities) to the GDP of the West region are 64.2% (Inner Mongolia 18.5%, Sichuan 20.4%, Shaanxi 13.2%, and Guangxi 12.1%). This is for the period of 2010–2020. For 2020–2030, the percentage total contributions of these four provinces will be 64.8% (Inner Mongolia 17.5%, Sichuan 20.9%, Shaanxi 14.2%, and Guangxi 12.2%).

As shown in Figure 5.9, for the next 20 years, the percentage GDP of the West region as a proportion of the national GDP will rise from 18.6% in 2010 to 19.5%

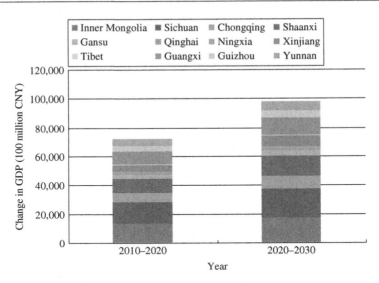

Figure 5.8 GDP changes of the various provinces (cities) in the West region.

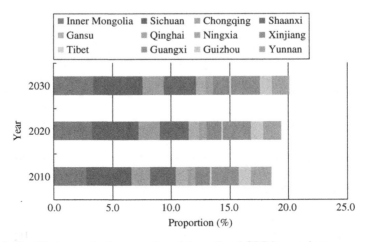

Figure 5.9 The Western region's proportion of the national GDP by provinces.

in 2020. By 2030, this will further increase to 20.1%. From the period during 2010–2030, the increase is almost 1.5 percentage points. From the level of the individual provinces (areas, cities), the percentage rise as a proportion of the national GDP rise for Inner Mongolia, Sichuan, Chongqing, and Shaanxi is more significant. The increases are 0.50, 0.16, 0.17, and 0.25 percentage points, respectively. For the 2020–2030 period, the percentage rise as a proportion of the national GDP will be

Table 5.18 Changes in the Industrial Sectors of the West Region (%)

Province (City)	Industry	2010	2020	2030
Inner Mongolia	Primary	8.7	3.7	2.5
	Secondary	56.1	52.4	45.3
	Tertiary	35.2	43.9	52.2
Sichuan	Primary	17.1	12.1	10.4
	Secondary	44.5	38.3	31.7
	Tertiary	38.4	49.6	57.9
Chongqing	Primary	9.4	4.5	3.2
	Secondary	45.1	42.5	39.1
	Tertiary	45.5	53.0	57.7
Shaanxi	Primary	10.1	8.0	7.1
	Secondary	53.0	44.4	38.2
	Tertiary	36.9	47.6	54.7
Gansu	Primary	12.2	8.6	7.7
	Secondary	46.7	40.1	34.1
	Tertiary	41.1	51.3	58.2
Qinghai	Primary	9.2	6.2	6.1
	Secondary	55.8	55.0	45.2
	Tertiary	35.0	38.8	48.7
Ningxia	Primary	9.6	7.1	6.6
	Secondary	50.4	44.3	37.0
	Tertiary	40.0	48.6	56.4
Xinjiang	Primary	14.9	13.2	12.8
	Secondary	49.3	38.3	26.3
	Tertiary	35.8	48.5	60.9
Tibet	Primary	13.3	6.3	4.8
	Secondary	32.1	42.1	42.6
	Tertiary	54.6	51.6	52.6
Guangxi	Primary	18.0	13.0	11.0
	Secondary	41.6	40.1	37.3
	Tertiary	40.4	46.9	51.7
Guizhou	Primary	14.1	9.9	8.8
	Secondary	41.6	37.3	32.9
	Tertiary	44.3	52.8	58.3
Yunnan	Primary	15.9	11.1	9.9
	Secondary	41.7	39.9	39.0
	Tertiary	42.4	49.0	51.1

significantly slowed down for Inner Mongolia and Chongqing, while for Sichuan and Shaanxi, it will still be quite large.

Table 5.18 lists out the structural changes in the 3-tier industry of the individual provinces (region, city) of the West region. Among the four major regions of the country, the level of development for the West region is the lowest and the initial proportion of its primary industry is the highest, reaching 14.5%, way above the national average by 4.3 percentage points, and is higher than the Central region by

Table 5.19 Simulated GDP Results of the Provinces in the Northeast Region (at Constant Price of 2005, per 100 million Yuan)

Individual Provinces in the Northeast Region	2010	2020	2030	2010–2020 Growth Rate (%)	2020–2030 Growth Rate (%)
Liaoning	12,558	29,641	54,029	9.0	6.2
Jilin	6118	14,012	24,463	8.6	5.7
Heilongjiang	7522	15,178	26,140	7.3	5.6
Northeast	26,198	58,831	104,632	8.4	5.9

GDP is calculated according to 2005 prices.
The last two columns of the table show the average GDP growth rate of the various regions for 2010–2020 and 2020–2030.

1.2 percentage points. By 2030, the proportion of primary industries for Guangxi, Sichuan, and Xinjiang remain at above 10%. In 2010, with the exception of Tibet, for all the 11 provinces (areas, cities) of the West region, the proportion of their tertiary industry remains at below 50%. Among them, Guangxi, Chongqing, Guizhou, Yunnan, and Gansu have proportion of tertiary industry that fall within the 40–50% range. The others are spread out within the range of 33–40%. By 2020, only the proportion of tertiary industries Chongqing, Guizhou, and Gansu will cross the 50% threshold. By 2030, the only area or city with its tertiary industry being less than 50% of the total is Qinghai. By 2030, the tertiary industry of the West region will be less than the national average, while its secondary industry will be higher than the national average.

The Northeast Region

Under the push of the "Northeast regeneration" economic strategy, the economic growth of the Northeast region will be slightly faster than the national average for the next 10–20 years. (For the simulated results of the Northeast region economic growth, see Table 5.19.) For 2010–2020, the annual GDP growth rate of the Northeast area is 8.4%, an average growth rate of 5.9%. The growth rate of these two stages is higher than the national average by 0.3 percentage points.

From the provincial level of development, Liaoning will spare no efforts to develop and strengthen its nonpublic owned style of economy, further developing its coastal economy, the untapped construction sector, and creating bases for its steel materials, rock and agricultural sectors, and so on to boost and equip its manufacturing industry and strengthen regional economic cooperation and hasten construction of basic facilities such as the railroad and ports. Its economy for the 2010–2020 period shall see faster growth, with annual growth rates of up to about 9%. This rate will drop to about 6.2% for the period of 2020–2030. For Jilin province, it will continue to push for the upgrading and optimization of its economic structure, expanding support for the construction of new farms, basic industries, basic infrastructure and picking up the pace for the development of its service sector. Its economy for the 2010–2020 period shall see it maintain a faster growth rate. For the years

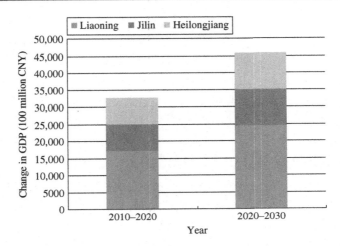

Figure 5.10 GDP changes of the various provinces (cities) in the Northeast region.

Table 5.20 Changes in the Industrial Sectors of the Northeast Region (%)

Province	Industry	2010	2020	2030
Liaoning	Primary		5.9	4.5
	Secondary		42.0	31.4
	Tertiary	37.2	52.1	64.1
Jilin	Primary	12.5	8.4	7.6
	Secondary	45.9	37.8	29.5
	Tertiary	41.6	53.8	62.9
Heilongjiang	Primary	12.4	11.7	11.2
	Secondary	50.3	40.1	32.1
	Tertiary	37.3	48.2	56.7

2020–2030, this rate will taper off to a more stable growth. Heilongjiang will greatly expand on its timber industry, wood processing industry, and the forest tourism industry, hastening the integrated exploration plus effective utilization of its coal mines, as well as its development of its other noncoal industries such as green foods, gold refinery, new age construction material, pharmaceuticals, and logistics.

From the changes in the GDP of the various provinces of the Northeast region, the main source of economic growth in the region would be Liaoning for the next 20 years. For 2010–2020, the contributions of Liaoning, Jilin, and Heilongjiang toward the GDP of the entire Northeast region are 52.3%, 24.2%, and 23.5%, respectively; for 2020–2030, the contributions of the three provinces are 53.3%, 22.8%, and 23.9%, respectively (Figure 5.10).

For the next 20 years, the proportion of the nation's GDP held by individual provinces of the Northeast region will rise from 8.6% in 2010 to 9.3% in 2030—a total of 0.7 percentage points from the performance of the individual provinces, between

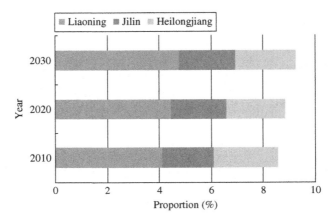

Figure 5.11 Proportion of the national GDP by provinces in Northeast region.

2010 and 2030, the proportion of the nation's GDP from Liaoning and Jilin will have risen by 0.7 and 0.2 percentage points, respectively, while Heilongjiang's proportion would drop by 0.2 percentage points (Figure 5.11).

Table 5.20 Lists out the changes in the 3-tier industrial sector structure of the individual provinces in the Northeast region. The initial proportion of primary and secondary industries for the Northeast region is higher than the national average. During the period of simulation, for Liaoning and Jilin in 2030, its proportion of tertiary industry is above 60% and for Heilongjiang, its proportion of tertiary industry is above 50%. By 2030, the average proportion of tertiary industry for the Northeast region is 62%, higher than the national average. A more special phenomenon is the decrease of Heilongjiang's primary industry, which is very small. This has to do with the well-developed agricultural sector of the particular province, which enjoys a significant comparative advantage nationally.

6 Scenario Analysis of China's Electricity Demand in the Year 2030

6.1 Scenario Analysis of the National Electricity Demand

6.1.1 Simulation Results for Scenario I

In Scenario I, the growth rate for total electricity demand is 6.16% between 2010 and 2020, which is lower than the GDP growth rate. The electricity elasticity for this time period will be 0.78. After 2020, China's electricity demand growth rate begins to slow significantly. Between 2020 and 2030, the growth rate for total electricity demand will be 2.55%, with an electricity elasticity of 0.47 (Table 6.1).

In this scenario, total electricity demand will reach 7.62 trillion kWh by 2020 and 9.81 trillion kWh by 2030 (Table 6.2). In 2020 and 2030, the electricity demand per capita will reach 5293 and 6672 kWh, respectively. The residential electricity demand per capita will reach 925 and 1448 kWh, respectively. The electricity intensity will decrease from 133 5 kWh/10,000 Yuan, in 2010, to 1135 kWh/10,000 Yuan in 2020, Finally, it will fall to 863 kWh/10,000 Yuan by 2030. From 2010 to2030, the electricity intensity will show an overall decrease of 35%.

According to the electricity demand structure (see Table 6.2), the proportion of electricity demand taken up by the primary and secondary industries will decrease, while the proportion taken up by tertiary industry and residential demand will increase. In 2020, the proportion of electricity demand occupied by the three industries and residential will be 1.9:67.9:12.7:17.5. When they are compared with 2010 figures, the primary and secondary industry proportions will decrease by 0.4 and 6.8 percentage points, respectively. In contrast to this, the tertiary industry and residential demand proportions will increase by 2.0 and 5.2 percentage points. By 2030, the proportions of electricity demand held by three industries and residential will be 1.8:61.3:15.3:21.7. When this is compared with 2020, the primary and secondary industry proportions will have decreased by 0.2 and 6.6 percentage points, and the tertiary industry and residential electricity demand proportions will have increased by 2.6 and 4.2 percentage points.

6.1.2 Simulation Results for Scenario II

In Scenario II, the growth rate for electricity demand is slightly higher than in Scenario I. For the period during 2010–2020, the total electricity demand growth rate will be 6.31%, which is higher than Scenario I by 0.15 percentage points. From 2020 to 2030, the growth rate will be 2.44%, which is lower than Scenario I by 0.11

An Exploration into China's Economic Development and Electricity Demand by the Year 2050.
DOI: http://dx.doi.org/10.1016/B978-0-12-420159-0.00006-0

Table 6.1 Total Electricity Demand Growth Rates and
Electricity Elasticity (%)

Index	2010–2020	2020–2030
Electricity demand growth rate	6.16	2.55
Electricity elasticity	0.78	0.47

Table 6.2 Future Levels of Total Annual National Electricity Demand

Index		2010	2020	2030
Electricity demand	Total	41,923	76,220	98,084
(100 million kWh)	Primary industry	984	1476	1718
	Secondary industry	31,318	51,737	60,098
	Tertiary industry	4497	9685	14978
	Residential	5125	13,322	21,290
Electricity demand	Primary industry	2.35	1.94	1.75
structure (%)	Secondary industry	74.70	67.88	61.27
	Tertiary industry	10.73	12.71	15.27
	Residential	12.22	17.47	21.71
Electricity demand per capita (kWh)		3126	5293	6672
Per capita residential electricity demand (kWh)		382	925	1448
Electricity intensity (kWh/10,000 Yuan, in constant 2005 Yuan)		1335	1135	863

Table 6.3 Total Electricity Demand Growth Rates and
Electricity Elasticity (%)

Index	2010–2020	2020–2030
Electricity demand growth rate	6.31	2.44
Electricity elasticity	0.78	0.44

percentage points. The electricity elasticity from 2010 to 2020 and 2020 to 2030 are 0.78 and 0.44, respectively (Table 6.3).

Considering the extent of electricity demand in Scenario II (Table 6.4), total electricity demand should reach 7.73 trillion kWh by 2020 and 9.84 trillion kWh by 2030. These figures are higher than Scenario I by 1.5% and 0.4%, respectively. In 2020 and 2030, electricity demand per capita will reach 5371 and 6697 kWh, residential electricity demand per capita will reach 925 kWh in 2020, and 1448 kWh in 2030. The electricity intensity will decrease from 1335 kWh/10,000 Yuan in 2010 to 1130 kWh/10,000 Yuan in 2020. It will further decrease to 834 kWh/10,000 Yuan by 2030. This will be a cumulative decrease of 37% over the 2010–2030, which is

Table 6.4 Future Levels of National Annual Electricity Demand

Index		2010	2020	2030
Electricity demand	Total	41,923	77,337	98,448
(100 million kWh)	Primary industry	984	1575	1851
	Secondary industry	31,318	51,137	58,059
	Tertiary industry	4497	11,303	17,247
	Residential	5125	13,322	21,290
Electricity demand	Primary industry	2.35	2.04	1.88
structure (%)	Secondary industry	74.70	66.12	58.97
	Tertiary industry	10.73	14.62	17.52
	Residential	12.22	17.22	21.63
Electricity demand per capita (kWh)		3126	5371	6697
Per capita residential electricity demand (kWh)		382	925	1448
Electricity intensity (kWh/10,000 Yuan, in constant 2005 Yuan)		1335	1130	834

higher than the reductions in Scenario I by 2 percentage points. In Scenario II, the situation where the growth of electricity demand relies on high energy intensity will change. At the same time, due to technological advancements, the efficiency of electricity demand will increase.

Considering the electricity demand structure (see Table 6.4), the proportion of electricity demand occupied by the secondary industry will be lower in Scenario II than Scenario I by 1.8~2.3 percentage points. However, the tertiary industry proportion will be higher than in Scenario I by 1.9~2.2 percentage points. Primary industry and residential electricity demand will be equal to Scenario I numbers. In 2020, the structure of three industries and residential electricity demand will be 2.0:66.1:14.6:17.2. When these figures are compared to 2010, the proportion of electricity demand of primary and secondary industries will decrease by 0.3 and 8.6 percentage points, respectively and the electricity demand of the tertiary industry and residential electricity demand will increase by 3.9 and 5.0 percentage points, respectively. In 2030, the structure of three industries and residential electricity demand will be 1.9:59.0:17.5:21.6. When compared with 2020, both the primary and the secondary industry proportions will decrease by 0.2 and 7.1 percentage points, respectively, while demand proportions by the tertiary industry and residential electricity demand will increase by 2.9 and 4.4 percentage points.

6.1.3 Simulation Results for Scenario III

In Scenario III, the electricity demand growth rate will be slightly lower than in Scenario I. From 2010 to 2020, the growth rate in total electricity demand will be 5.86%, which is lower than in Scenario I by 0.3 percentage points. From 2020 to 2030, the growth rate will be 2.50%. The electricity elasticity are 0.78 from 2010 to 2020 and 0.51 from 2020 to 2030 (Table 6.5).

Table 6.5 National Electricity Demand Growth Rates
and Electricity Elasticity (%)

Index	2010–2020	2020–2030
Electricity demand growth rate	5.86	2.50
Electricity elasticity	0.78	0.51

Table 6.6 Future Levels of National Annual Electricity Demand

Index		2010	2020	2030
Electricity demand	Total	41,923	74,067	94,843
(100 million kWh)	Primary industry	984	1495	1680
	Secondary industry	31,318	49,932	57,740
	Tertiary industry	4497	9319	14 133
	Residential	5125	13,322	21,290
Electricity demand	Primary industry	2.35	2.02	1.77
structure (%)	Secondary industry	74.70	67.41	60.88
	Tertiary industry	10.73	12.58	14.90
	Residential	12.22	17.99	22.45
Electricity demand per capita (kWh)		3126	5144	6452
Per capita residential electricity demand (kWh)		382	925	1448
Electricity intensity (kWh/10,000 Yuan, in constant 2005 Yuan)		1335	1144	908

Considering the extent of electricity demand in Scenario III (Table 6.6), total electricity demand will reach 7.41 trillion kWh by 2020 and 9.48 trillion kWh by 2030. In 2020 and 2030, electricity demand per capita will be 5144 and 6452 kWh, respectively, and residential electricity demand per capita will reach 925 and 1448 kWh. Electricity intensity will decrease from 1335 kWh/10,000 Yuan in 2010 to 1144 kWh/10,000 Yuan in 2020 and 908 kWh/10,000 Yuan in 2030. This is a cumulative decrease of 32% over the 2010–2030, which is a degree of reduction that is lower than Scenario I by 3 percentage points.

According to the electricity demand structures (see Table 6.6), the electricity demand structure for the three industries and residential under Scenario III will be 2.0:67.4:12.6:18.0 in 2020. When this is compared with 2010, the primary and secondary industry proportions will have declined by 0.3 and 7.3 percentage points, while the tertiary industry and residential electricity demand proportions will have increased by 1.8 and 5.8 percentage points. In 2030, the three industries and residential electricity demand structure will be 1.8:60.9:14.9:22.5. When this is compared with 2020, the primary and secondary industry proportions will have decreased by 0.3 and 6.5 percentage points, and the tertiary industry and residential electricity demand proportions will have increased by 2.3 and 4.5 percentage points, respectively.

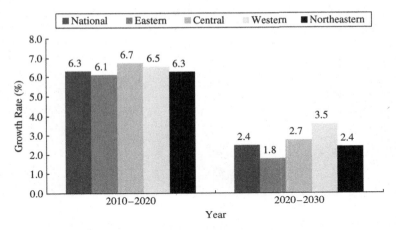

Figure 6.1 National and regional electricity demand growth rates for each decade from 2010 to 2030.

6.2   Scenario Analysis of the Regional Electricity Demand

Along with the scenario analysis results for the regional economic development, this section also describes the regional electricity demand in Scenario II.

6.2.1   Regional Comparisons

In the future, with the advantages of their labor and natural resources, the central and western regions will gradually carry on industry transfers and its growth rate of electricity demand will be smaller than the rate of eastern region.

Between 2010 and 2020, the growth rate in electricity demand for the central and western regions will be equal. These two regions have the highest electricity demand growth rates in the nation and are followed closely by the northeast. The east has the lowest electricity demand growth rate of all regions. From 2020 to 2030, all regions will experience significant decreases in their electricity demand growth rates. The most significant decrease will be in the eastern region, which will also continue to be the region which shows the slowest rate of growth. During this period, the growth rate of electricity demand in the west will exceed that of the central region. Furthermore, the northeastern region will continue to maintain a growth rate that is lower than both the western and the central regions. The national and regional electricity demand growth rates for each decade from 2010 to 2030 are shown in Figure 6.1.

Considering the regional proportions of electricity demand (Figure 6.2), demand in the east occupies a higher proportion of national demand than does the three other regions. The western region comes second, and the central region ranks third. The northeast occupies the lowest proportion of national electricity demand. In the future, the central and western regions will make up an increasing proportion of

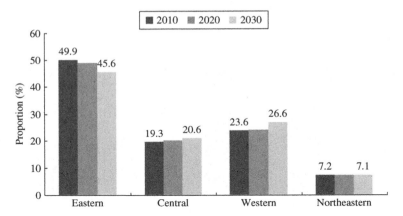

Figure 6.2 Regional proportions of electricity demand from 2010 to 2030.

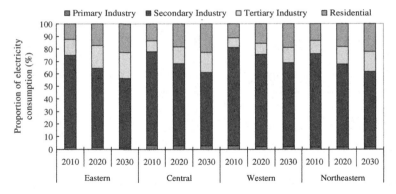

Figure 6.3 Changes in the regional electricity demand structure from 2010 to 2030.

national electricity demand, with the most significant increases in the west. The proportion of national electricity demand occupied by the eastern region will progressively decline, while changes in the northeastern proportion will be minor.

Considering the electricity demand structures (Figure 6.3), each of the four regions show continued decreases for the proportions of electricity demand held by the primary and secondary industries. The tertiary industry and residential electricity demand proportions will continue to increase. The regional primary industry's proportion will decrease by less than 1%, and the secondary industry's proportion will decrease significantly. The regional proportion of electricity demand occupied by residential demand will rise more than that of the tertiary industry.

In 2030, the proportion of electricity demand held by secondary industry will be comparatively high in the western region, while the tertiary industry's proportion will be relatively low. In the east, the lowest demand proportion will be taken by secondary industry, while tertiary industry will occupy the greatest proportion. In the

Table 6.7 Total Electricity Demand in the Eastern Region (100 million kWh)

Eastern Region and Provinces (Cities)	2010	2020	2030	2010–2020 Growth Rate (%)	2020–2030 Growth Rate (%)
Beijing	800	1199	1332	4.1	1.1
Tianjin	643	1109	1367	5.6	2.1
Hebei	2692	5424	6769	7.3	2.2
Shandong	3300	6069	7603	6.3	2.3
Shanghai	1290	2071	2328	4.8	1.2
Jiangsu	3856	6920	7915	6.0	1.4
Zhejiang	2825	5131	5968	6.2	1.5
Fujian	1316	2463	3085	6.5	2.3
Guangdong	4060	7025	8151	5.6	1.5
Hainan	158	345	474	8.1	3.2
All Eastern	20,939	37,756	44,993	6.1	1.8

The last two columns in the table represent the growth rates in total electricity demand for all of the regions for the periods of 2010–2020 and 2020–2030.

central and northeastern regions, the secondary industry proportion will be slightly higher than national average levels, while the tertiary industry's proportion will be lower than the national average. In the central and northeastern regions, residential will take up a larger proportion of electricity demand than the national average. In the eastern region, this proportion will be equal to the national average. In the western region, the proportion will be significantly lower than the national average.

6.2.2 The Eastern Region

In the eastern region, the growth rate for electricity demand will show substantial decreases over the next 10–20 years, and it will also be significantly lower than the national average (simulation results for electricity demand in the eastern region are shown in Table 6.7). Between 2010 and 2020, in the eastern region, the average annual growth rate for electricity demand will be 6.1%, which is lower than the national average by 0.2 percentage points. Between 2020 and 2030, the average growth rate in electricity demand here will be 1.8%, which is lower than the national average by 0.6 percentage points.

From 2010 to 2020, in the eastern region, Hebei, Fujian, and Hainan will have electricity demand growth rates that are slightly higher than the national average, within a range between 6.5% and 8.1%. The growth rate in Shandong will equal the national average. The other six provinces (cities) will have electricity demand growth rates that are lower than the national average, with rates between 4.1% and 6.2%. Among these, Beijing's growth rate will be 4.1%, lower than the national average by 2.2 percentage points. Shanghai will have a growth rate of 4.8%, which is 1.5 percentage points lower than the national average.

Between 2020 and 2030 in the eastern region, only Hainan will have a growth in electricity demand that is higher than the national average. Here, it will reach 3.2%.

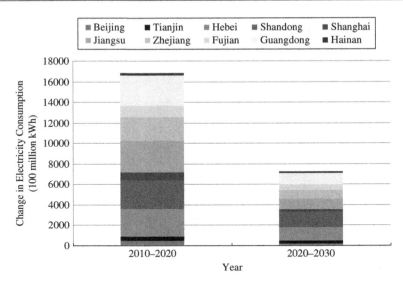

Figure 6.4 Changes in electricity demand in the eastern provinces.

The four municipal provinces of Tianjin, Hebei, Shandong, and Fujin will have growth rates slightly lower than the national average by about 0.1~0.2 percentage points. Beijing, Shanghai, Jiangsu, Zhejiang, and Guangdong will have growth rates that are significantly lower (2%) than the national average levels. This demonstrates that, within these provinces (cities), electricity demand growth rates will become saturated before 2030.

According to the changes in electricity demand in the eastern provinces (cities), this region will experience increases in total electricity demand which are primarily due to demand from the provinces of Guangdong, Jiangsu, Shandong, and Hebei (Figure 6.4). When compared with 2010, the 2020 growth rate in overall electricity demand in the eastern region will increase by 1.68 trillion kWh. The accumulated contribution rate to electricity demand by the four aforementioned provinces will be 68.5% (Guangdong 17.6%, Jiangsu 18.2%, Shandong 16.5%, Hebei 16.2%). From 2020 to 2030, overall electricity demand in the eastern region will grow by 0.72 trillion kWh. The contributions from the above four provinces for growth in electricity demand will be 69.2% (Guangdong 15.6%, Jiangsu 13.8%, Shandong 21.2%, Hebei 18.6%).

Along with the substantial reductions in the electricity demand growth rate, the proportion of national electricity demand held by the eastern region will decrease from 49.9% in 2010 to 48.8% in 2020, and further to 45.7% in 2030. This trend is shown in Figure 6.5. This is a decline of 4.2 percentage points over the 2010–2030 period, with an average annual decline of 0.21 percentage points. Among the provinces (cities), from 2010 to 2020, Hebei, Fujian, and Hainan will demonstrate a somewhat increased proportion of electricity demand. The degree of increase in Hebei will be comparatively large, reaching 0.6 percentage points. In Shandong, the

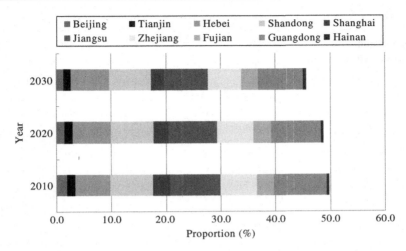

Figure 6.5 The proportion of national electricity demand held by eastern provinces.

proportion will remain basically unchanged. Beijing, Tianjin, Shanghai, Jiangsu, Zhejiang, and Guangdong, will experience decreases in their proportions of electricity demand. Of those six provinces, Guangdong will show substantially large decreases, which will reach 0.6 percentage points. From 2020 to 2030, electricity demand in Hainan will occupy a slightly increased proportion of the national demand. In all of the other nine provinces (cities), the proportions of national electricity demand will decrease. Of these provinces, Jiangsu, Zhejiang, and Guangdong will show decreases of 0.9, 0.6, and 0.8 percentage points, respectively.

According to the structural changes in electricity demand (Table 6.8), from 2010 to 2030, the proportion of electricity demand occupied by primary industry in the eastern region will decrease by 0.5 percentage points. Secondary industry will decrease by 18.5 percentage points, and tertiary industry and residential electricity demand will increase by 8.8 and 10.2 percentage points, respectively.

In the eastern region, the proportion of electricity demand held by the primary and secondary industries will gradually decrease in all provinces (cities), while the tertiary industry and residential electricity demand proportions will gradually increase. From 2020, the secondary industry proportion in the other provinces (cities) will all be 50% or higher (this excludes the secondary industry proportions in Beijing and Shanghai which will be lower than 41%). Of these, the secondary industry demand proportions in Hebei, Shandong, Jiangsu, and Zhejiang will all be relatively high, at around 70%. The proportions for Fujian, Guangdong, and Hainan will be relatively low, at 60.3%, 54.4%, and 50.1%, respectively. In 2030, the proportions of electricity demand held by secondary industry in Beijing, Tianjin, Shanghai, Zhejiang, Fujian, Guangdong, and Hainan will all be lower than the national average. The proportions in Hebei, Shandong, and Jiangsu will have remained relatively high, in the ranges of 60.0–64.0%.

Table 6.8 Structural Changes in Electricity Demand in the Eastern Region (%)

Eastern Region and Provinces (Cities)	Sector	2010	2020	2030
Whole eastern region	Primary industry	1.9	1.6	1.4
	Secondary industry	73.1	62.9	54.6
	Tertiary industry	12.7	18.2	21.5
	Residential	12.3	17.3	22.5
Beijing	Primary industry	2.1	1.2	0.8
	Secondary industry	40.5	25.5	18.6
	Tertiary industry	40.2	45.7	48.2
	Residential	17.2	27.6	32.4
Tianjin	Primary industry	1.8	1.0	0.5
	Secondary industry	73.6	59.8	53.9
	Tertiary industry	14.2	21.9	25.4
	Residential	10.4	17.3	20.2
Hebei	Primary industry	5.9	5.6	5.0
	Secondary industry	76.6	68.2	61.3
	Tertiary industry	7.0	11.7	14.3
	Residential	10.5	14.5	19.4
Shandong	Primary industry	2.5	1.6	1.4
	Secondary industry	78.3	69.5	60.6
	Tertiary industry	8.0	13.1	16.8
	Residential	11.2	15.8	21.2
Shanghai	Primary industry	0.5	0.3	0.2
	Secondary industry	61.5	40.9	31.6
	Tertiary industry	25.0	38.7	44.8
	Residential	13.0	20.1	23.4
Jiangsu	Primary industry	0.8	0.6	0.5
	Secondary industry	80.3	73.3	63.8
	Tertiary industry	9.2	12.7	16.3
	Residential	9.7	13.4	19.4
Zhejiang	Primary industry	0.6	0.4	0.3
	Secondary industry	78.2	67.5	58.1
	Tertiary industry	9.9	15.2	19.4
	Residential	11.3	16.9	22.2
Fujian	Primary industry	1.0	0.8	0.6
	Secondary industry	69.5	60.3	54.2
	Tertiary industry	11.4	15.2	16.8
	Residential	18.1	23.7	28.4
Guangdong	Primary industry	1.4	1.0	0.8
	Secondary industry	68.0	54.4	45.2
	Tertiary industry	16.1	24.2	27.8
	Residential	14.5	20.4	26.2
Hainan	Primary industry	4.3	3.3	2.9
	Secondary industry	56.9	50.1	46.8
	Tertiary industry	25.2	30.4	33.0
	Residential	13.6	16.2	17.3

Table 6.9 Growth in Total Electricity Demand in the Central Region (100 million kWh)

Central Region and Provinces	2010	2020	2030	2010–2020 Growth Rate (%)	2020–2030 Growth Rate (%)
Shanxi	1450	3102	3703	7.9	1.8
Anhui	1076	2011	2400	6.5	1.8
Henan	2355	4583	6142	6.9	3.0
Hubei	1325	2431	3417	6.3	3.5
Hunan	1177	2166	2987	6.3	3.3
Jiangxi	700	1199	1653	5.5	3.3
The central region	8083	15,491	20,302	6.7	2.7

The last two columns in the table represent the growth rates in total regional electricity demand from the periods of 2010–2020 and 2020–2030.

6.2.3 The Central Region

In the central region, the growth rate in electricity demand will show substantial decreases over the next 10–20 years, but the growth rate will still be higher than the national average (central region electricity demand simulation results are shown in Figure 6.9). From 2010 to 2020, the average annual growth rate for electricity demand in the central region will be 6.7%. This figure is higher than the national average by 0.4 percentage points. From 2020 to 2030, the growth rate will be 2.7%, higher than the national average by 0.3 percentage points.

Between 2010 and 2020, in the central region, Shanxi, Anhui, and Henan will have electricity demand growth rates that are higher than the national average. They will range between 6.5% and 7.9%. Among these, electricity demand in Shanxi will increase relatively quickly, with a growth rate reaching 7.9%, which is higher than the national average by 1.6 percentage points. Hubei and Hunan will have growth rates about the same as the national average. In Jiangxi, the growth in electricity demand will be lowest, with a growth rate of only 5.5%.

From 2020 to 2030, Shanxi and Anhui provinces will have electricity demand growth rates that show relatively rapid decreases, and which are lower than the national average. The other four provinces will have higher growth rates than the national average. The provincial growth rate is relatively balanced and shows no large discrepancies between the provinces. The electricity demand growth rate is about 3.3% (Table 6.9).

According to provincial electricity demand changes in the central region (Figure 6.6), from 2010 to 2020, increases in total electricity demand will mainly come from Henan and Shanxi. During this period, these two provinces will contribute 52.4% (Henan 30.1%, Shanxi 22.3%) of the growth in electricity demand for the region. From 2020 to 2030, growth in Shanxi will be relatively slow, with a substantially decreased contribution rate to the region's growth in electricity demand. Conversely, Hubei and Henan will show substantially increased contributions to the growth in electricity demand. The main sources of growth in the region will come from Henan, Hubei, and Hunan. During this period, these three provinces will have

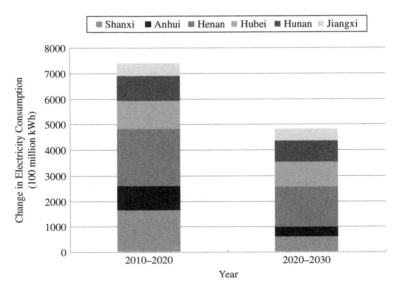

Figure 6.6 Changes in electricity demand in the central provinces.

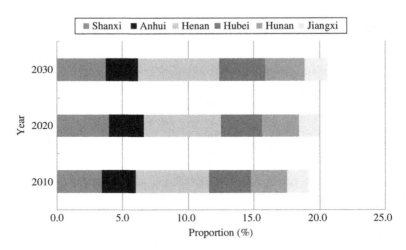

Figure 6.7 Proportion of national electricity demand held by central provinces.

a contribution rate of 70.0% to the growth in electricity demand for the entire region (Henan 32.4%, Hubei 20.5%, and Hunan 17.1%).

In the central region over the next 20 years, due to electricity demand growth rates that are higher than the national average, the proportion of national demand held by this region will increase from 19.3% in 2010, to 20.0% in 2020, and further to 20.6% by 2030 (Figure 6.7). This will be a total increase of 1.3 percentage points

Table 6.10 Proportional Changes in Electricity Demand in the Central Region (%)

Central Region and Provinces (Cities)	Sector	2010	2020	2030
Whole central region	Primary industry	3.1	2.9	2.8
	Secondary industry	74.9	65.6	58.4
	Tertiary industry	8.9	13.0	16.0
	Residential	13.1	18.5	22.8
Shanxi	Primary industry	2.4	2.0	1.9
	Secondary industry	82.6	72.1	65.8
	Tertiary industry	7.8	14.6	16.4
	Residential	7.2	11.3	15.9
Anhui	Primary industry	1.1	0.9	0.6
	Secondary industry	74.0	67.8	60.4
	Tertiary industry	8.9	9.5	14.7
	Residential	16.0	21.8	24.3
Henan	Primary industry	3.3	2.9	2.7
	Secondary industry	77.7	69.7	62.1
	Tertiary industry	7.5	12.9	15.7
	Residential	11.5	14.5	19.5
Hubei	Primary industry	1.5	1.4	1.2
	Secondary industry	73.5	61.4	54.9
	Tertiary industry	10.4	14.1	17.1
	Residential	14.6	23.1	26.8
Hunan	Primary industry	8.2	8.2	7.8
	Secondary industry	64.1	53.6	46.8
	Tertiary industry	10.3	12.3	15.3
	Residential	17.4	25.9	30.1
Jiangxi	Primary industry	1.8	2.6	2.3
	Secondary industry	71.9	59.9	53.5
	Tertiary industry	10.5	13.5	16.8
	Residential	15.8	24.0	27.4

over the 2010–2030. In the central provinces from 2010 to 2020, Shanxi, Anhui, and Henan will occupy an increasing proportion of electricity demand. Shanxi and Henan will show the most significant increases at 0.6 and 0.3 percentage points, respectively. Hubei and Henan will show little or no change in their proportions of electricity demand, while Jiangxi will experience a slight decrease. From 2020 to 2030, electricity demand in Shanxi and Anhui will occupy a declining proportion of total demand by 0.25 and 0.16 percentage points, respectively. The remaining four provinces will show increased proportions. Of these, Henan and Hubei will have proportions that increase by 0.31 and 0.33 percentage points, respectively.

According to structural changes in electricity demand (Table 6.10), from 2010 to 2030, the primary industry's electricity demand proportion will decrease by 0.3 percentage points within the central region. The secondary industry's demand proportion will decrease by 16.5 percentage points, and proportions from the tertiary

Table 6.11 Growth of Total Electricity Demand in the Western Region (100 million kWh)

Western Region and Provinces (Regions, Cities)	2010	2020	2030	2010–2020 Growth Rate (%)	2020–2030 Growth Rate (%)
Inner Mongolia	1530	3185	4359	7.6	3.2
Sichuan	1550	2735	3665	5.8	3.0
Chongqing	626	1229	1711	7.0	3.4
Shaanxi	853	1556	2335	6.2	4.1
Gansu	803	1369	2051	5.5	4.1
Qinghai	465	781	969	5.3	2.2
Ningxia	547	977	1413	6.0	3.8
Xinjiang	648	1410	2441	8.1	5.6
Tibet	20	46	73	8.4	4.7
Guangxi	993	1898	2543	6.7	3.0
Guizhou	836	1544	2180	6.3	3.5
Yunnan	1000	1805	2389	6.1	2.8
All Western	9872	18,533	26,129	6.5	3.5

The last two columns in the table represent the regional growth rates for total electricity demand for the periods of 2010–2020 and 2020–2030.

industry and residential electricity demand will increase by 7.1 and 9.7 percentage points, respectively.

In all of the provinces, the proportion of electricity demand occupied by the primary and secondary industries will gradually decrease. The tertiary industry and residential electricity demand proportions will gradually increase. By 2020, the proportion of electricity demand occupied by the secondary industry in Shanxi and Henan will be about 70%, a figure which is significantly higher than the other four provinces. Anhui, Hebei, Henan, and Jiangxi will have relatively high residential electricity demand proportions, all of which surpass 20%. These figures will exceed the tertiary industry's electricity demand proportion by 10 percentage points. By 2030, Shanxi and Henan will still have secondary industry demand proportions higher than 62%. The proportion in Anhui will be about 60.4%, and, for the other three provinces, the proportion will be less than 55%.

6.2.4 The Western Region

Of the four main regions, the growth rate for electricity demand will be highest in the western region (electricity demand simulation results for the western region are shown in Table 6.11). From 2010 to 2020, the average annual growth rate in electricity demand for the western region will be 6.5%. This will be higher than the national average by about 0.2 percentage points. From 2020 to 2030, the growth rate will be 3.5%, or higher than the national average by about 1.1 percentage points.

From 2010 to 2020, Inner Mongolia, Chongqing, Xinjiang, Tibet, and Guangxi will have electricity demand growth rates higher than the national average. All will

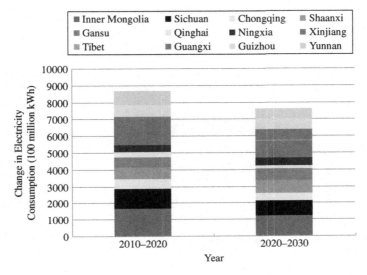

Figure 6.8 Changes in electricity demand in the western provinces (region, city).

be between 6.7% and 8.4%. Shaanxi and Guizhou will have growth rates that equal the national average. Sichuan, Gansu, Qinghai, Ningxia, and Yunnan will all have growth rates that are lower than the national average, between 5.3% and 6.1%.

From 2020 to 2030, the western provinces (region, cities) will still show relatively large discrepancies in their electricity demand growth rates. The rates will vary between 2.2% and 5.6%. The growth rate will be between 2% and 3% for Qinghai, Guangxi, and Yunnan. There will be a 3~4% growth rate for Inner Mongolia, Sichuan, Chongqing, Guangxi, Guizhou, and a 4~5% growth rate for Shaanxi, Gansu, and Tibet. Only Xinjiang will have a growth rate that exceeds 5%.

According to changes in electricity demand between 2010 and 2020 for the western provinces (regions, cities) (Figure 6.8), total electricity demand in the west will increase mainly due to increase demand in Inner Mongolia, Sichuan, Guangxi, and Yunnan.

During this period, the four provinces (regions) will have an accumulated contribution rate of 52.6% (Inner Mongolia 19.1%, Sichuan 13.7%, Guangxi 10.5%, Yunnan 9.3%). From 2020 to 2030, growth in electricity demand will be relatively slow in Guangxi and Yunnan, with each showing a substantially decreased contribution rate to growth in electricity demand. Shaanxi and Xinjiang will both show relatively rapid growth, with a substantially increased contribution rate to electricity demand growth for the region. Electricity demand growth in the western region will mainly be driven by Inner Mongolia, Sichuan, Shaanxi, and Xinjiang (region). During this period, these four provinces (regions) will contribute 51.6% to the region's growth (Inner Mongolia 15.5%, Sichuan 12.2%, Shaanxi 10.3%, Xinjiang 13.6%).

Over the next 20 years, due to the electricity demand growth rate is continuously higher than the national average; the western region will make up an increased

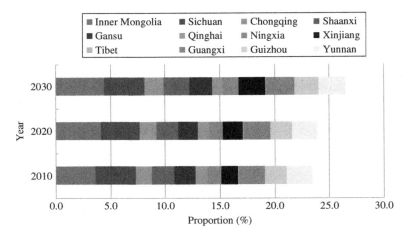

Figure 6.9 Proportion of national electricity demand held by the western provinces (region, city).

proportion of electricity demand. These proportions will rise from 23.6% in 2010 to 24% in 2020, and further to 26.6% by 2030 (Figure 6.9). This will be an increase of nearly 3 percentage points over the 2010–2030. For the provinces (regions, cities), Inner Mongolia and Xinjiang will show some relatively significant increases in their proportions of electricity demand from 2010 to 2020, with growth of 0.47 and 0.28 percentage points, respectively. From 2020 to 2030, aside from the 0.03 percentage point decrease in the Qinghai electricity demand proportion, all of the other western provinces (regions, cities) will take up an increased proportion of electricity demand. Among these 11 provinces, Inner Mongolia, Shaanxi, Gansu, and Xinjiang will have proportions that increase from 0.3 to 0.7 percentage points.

According to the changes in electricity demand for the west (Table 6.12), from 2010 to 2030, the proportion of electricity demand occupied by primary industry will decrease by 0.6 percentage points. The secondary industry's proportion will decrease by 11.6 percentage points, and tertiary industry and residential electricity demand proportions will increase by 4.3 and 7.9 percentage points, respectively.

The western region has a high concentration of energy intensive provinces. In the majority of these provinces (region and city), the proportion of electricity demand by secondary industry is significantly higher than the national average. Other than Tibet, the main structural changes in electricity demand for the western provinces (region and city) will be decreased in demand by the primary and secondary industries. The tertiary industry and residential electricity demand proportions will both show increases. In Tibet, the secondary industry's demand proportion will continue to increase through the year 2030.

By 2020, the proportion of electricity demand occupied by the secondary industry in Inner Mongolia, Gansu, Qinghai, Ningxia, Xinjiang, Guangxi, Guizhou, and Yunnan will be higher than the national average. Among these provinces, secondary industry's demand in Inner Mongolia will reach 84.5%, and Qinghai and Ningxia

Table 6.12 Changes in the Proportion of Electricity Demand in the Western Region (%)

Western Region and Provinces (Regions, Cities)	Sector	2010	2020	2030
Whole western region	Primary industry	2.8	2.3	2.2
	Secondary industry	78.1	73.1	66.5
	Tertiary industry	8.0	9.0	12.3
	Residential	11.1	15.6	19.0
Inner Mongolia	Primary industry	2.6	1.9	1.5
	Secondary industry	87.9	84.5	77.2
	Tertiary industry	4.5	6.1	10.2
	Residential	5.0	7.5	11.1
Sichuan	Primary industry	0.9	0.8	0.7
	Secondary industry	73.8	63.3	54.6
	Tertiary industry	9.7	11.4	17.5
	Residential	15.6	24.5	27.2
Chongqing	Primary industry	0.2	0.1	0.1
	Secondary industry	69.4	58.2	51.3
	Tertiary industry	14.1	14.3	16.7
	Residential	16.3	27.4	31.9
Shaanxi	Primary industry	5.1	4.3	3.9
	Secondary industry	68.2	62.3	53.2
	Tertiary industry	14.8	16.8	20.8
	Residential	11.9	16.6	22.1
Gansu	Primary industry	6.7	5.9	5.3
	Secondary industry	78.0	75.3	67.6
	Tertiary industry	8.7	9.3	12.4
	Residential	6.6	9.5	14.7
Qinghai	Primary industry	0.4	0.2	0.2
	Secondary Industry	93.7	90.4	78.1
	Tertiary industry	2.9	4.6	10.2
	Residential	3.0	4.8	11.5
Ningxia	Primary industry	2.3	1.9	1.6
	Secondary industry	91.2	90.1	79.3
	Tertiary industry	3.2	3.5	8.7
	Residential	3.3	4.5	10.4
Xinjiang	Primary industry	10.4	8.8	7.2
	Secondary industry	73.8	73.4	70.2
	Tertiary industry	8.2	8.6	10.1
	Residential	7.6	9.2	12.5
Tibet	Primary industry	1.9	2.3	2.6
	Secondary industry	46.6	50.6	55.5
	Tertiary industry	22.2	25.3	28.8
	Residential	29.3	21.8	13.1
Guangxi	Primary industry	2.1	1.8	1.6
	Secondary industry	74.2	71.2	67.6
	Tertiary industry	7.9	8.7	11.1
	Residential	15.8	18.3	19.7

(*Continued*)

Table 6.12 (Continued)

Western Region and Provinces (Regions, Cities)	Sector	2010	2020	2030
Guizhou	Primary industry	0.5	0.5	0.4
	Secondary industry	77.9	71.7	68.6
	Tertiary industry	6.9	7.8	8.3
	Residential	14.7	20.0	22.7
Yunnan	Primary industry	1.1	0.9	0.8
	Secondary industry	77.1	72.2	69.8
	Tertiary industry	6.2	5.9	5.8
	Residential	15.6	21.0	23.6

Table 6.13 Growth in Total Electricity Demand in the Northeastern Region (100 million kWh)

Northeastern Region and Provinces	2010	2020	2030	2010–2020 Growth Rate (%)	2020–2030 Growth Rate (%)
Liaoning	1717	3174	4093	6.34	2.58
Jilin	572	1032	1397	6.08	3.07
Heilongjiang	740	1350	1535	6.19	1.29
All Northeastern	3029	5557	7025	6.25	2.37

The last two columns in the table represent the regional growth rates for total electricity demand from the periods of 2010–2020 and 2020–2030.

will reach 90.4% and 90.1%, respectively. By 2030, the electricity demand proportion by secondary industry in most western provinces (regions, cities) will remain relatively high. Ningxia will reach 79.3%, and Qinghai and Inner Mongolia will reach 78.1% and 77.2%, respectively. Gansu, Xinjiang, Guangxi, Guizhou, and Yunnan will have proportions of between 67% and 71%. Secondary industry in Sichuan, Chongqing, Shaanxi, and Tibet will have relatively low proportions, all less than 56%.

6.2.5   The Northeastern Region

The results of electricity demand simulation for the northeastern region are shown in Table 6.13. Over the next 10–20 years, growth in electricity demand within this region will remain relatively slow. The growth rate will be slightly lower than national average by 0.1 percentage points. From 2010 to 2020, the average annual growth rate for electricity demand in the region will be 6.25%, and it will drop to 2.37% from 2020 to 2030.

From 2010 to 2020, only Ningxia's electricity demand growth rate will be slightly higher than the national average. Jilin and Heilongjiang will have lower growth rates than the national average.

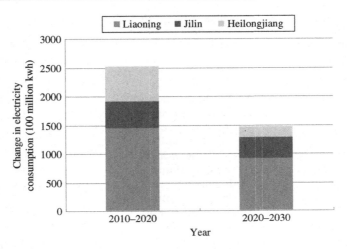

Figure 6.10 Changes in electricity demand in the northeastern provinces.

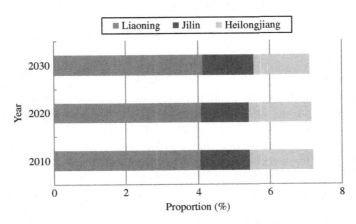

Figure 6.11 Proportion of national electricity demand held by northeastern provinces.

From 2020 to 2030, the electricity demand growth rate in Heilongjiang will be lower than the national average, and the growth rate in Ningxia and Jilin will be slightly higher than the national average.

According to the changes in electricity demand for the northeast (Figure 6.10), over the next 20 years, Liaoning will be the main source of growth in regional electricity demand. From 2010 to 2020, Liaoning, Jilin, and Heilongjiang will contribute 57.7%, 18.2%, and 24.1% to growth in the region, respectively. From 2020 to 2030, the three provinces will have contributions that reach 62.6%, 24.8%, and 12.6%, respectively.

Due to the relatively low electricity demand in the northeastern region, the proportion of national demand will show no significant changes (Figure 6.11). Over the next 20 years, the proportions will decrease from 7.2% in 2010 to 7.1% in 2030.

Table **6.14** Changes in the Proportion of Electricity Demand in Northeastern Region (%)

Northeastern Region and Provinces	Sector	2010	2020	2030
Whole northeastern region	Primary industry	1.9	1.7	1.5
	Secondary industry	74.0	66.2	60.4
	Tertiary industry	10.6	13.8	16.2
	Residential	13.5	18.3	21.9
Liaoning	Primary industry	1.3	1.1	1.0
	Secondary industry	77.7	71.6	65.7
	Tertiary industry	10.4	12.9	15.1
	Residential	10.6	14.4	18.2
Jilin	Primary industry	1.6	1.4	1.2
	Secondary industry	69.2	62.8	56.7
	Tertiary industry	11.9	15.1	17.8
	Residential	17.3	20.7	24.3
Heilongjiang	Primary industry	3.6	3.2	3.1
	Secondary industry	69.1	56.1	49.6
	Tertiary industry	10.2	15.0	17.5
	Residential	17.1	25.7	29.8

This will represent a total decrease of 0.1 percentage points. In terms of the provinces, from 2010 to 2030, Liaoning will account for a slightly increased proportion of electricity demand, and Jilin and Heilongjiang will have decreased proportions. From 2020 to 2030, the proportion of demand held by Liaoning will continue to increase slightly, the proportion held by Jilin will also slightly increase, and the proportion held by Heilongjiang will continue to decrease.

According to the structural changes in electricity demand for the northeast (Table 6.14), from 2010 to 2030, primary industry's electricity demand will decrease by 0.4 percentage points, regionally. Secondary industry's demand will decrease by 13.6 percentage points, and the demand from the tertiary industry and residential will increased by 5.6 and 8.4 percentage points, respectively.

The three provinces in the northeastern region will all show similar changes in their electricity demand structure. Each will show decreases in the proportions of electricity demand held by the primary and secondary industries, while the demand proportions from tertiary industry and residential will increase. In Heilongjiang, the residential electricity demand proportion will increase relatively quickly.

By 2020, the proportion of electricity demand held by the secondary industry in Liaoning will be higher than the national average, reaching 71.6%. In Jilin and Heilongjiang, the proportions will all be lower than the national average, at 62.8% and 56.1% respectively. By 2030, Liaoning, Jilin, and Heilongjiang will have secondary industry electricity demand proportions of 65.7%, 56.7%, and 49.6%, respectively. Jilin and Heilongjiang will have residential electricity demand proportions that reach 24.3% and 29.8%, respectively. This will be higher than the national average by 2.7 and 8.2 percentage points.

7 China's Economy and Electricity Demand Outlook for 2050

7.1 Model Constraints Setting

In ILE4, the Solow model and the input–output models are used for forecast in two steps: first, the use of the Solow model forecasts China's 2050 total GDP. On the basis of these figures, the input–output model can then be used to forecast economic structure and demand for electricity in 2050.

During the use of the Solow model to forecast the total GDP in 2050, historical data from 1952 to 2009 was utilized to determine the coefficients in the model. Among them, the GDP is based on calendar year GDP index in accordance with the findings derived from the price converted from 1952; labor supply data comes from the Statistical Yearbook of each year, the number of employees; the capital stock of the initial period of domestic scholars is set to 103 billion yuan (monetary value in 1952).[1,2,3] Given the depreciation rate of 8%, the fixed capital formation growth rate for total investment in fixed assets price index will be the calendar year, and gross fixed capital formation will be converted to 1952 prices. This is expressed by the net investment.

When we are using the input–output model for forecast, all the basic data applied are derived from China's input–output table in 2007.

During the forecasting, the exogenous variables given include future labor growth rate, the proportion of total fixed assets formation to the previous GDP, the total productivity growth rate, proportion of three major demands and the direct consumption coefficient, etc.

7.1.1 Labor Supply

The growth of the labor force depends on total population growth. It also depends on population age structure and the labor force participation rate. Due to the fact that this book lacks a detailed population model, it used the judgment of some institutions and scholars on China's labor force growth as references. Some research directly references the United Nations in terms of the labor supply forecasts for China's labor force aged population. Figure 7.1 is the United Nations forecast of the future population of the labor force aged 15~64 years.

Due to the fact that the supply of the working-age population is only a potential labor supply number, this part of population does not always directly participate in labor market activities. For this part of the population, those that really belong to the labor supply part of the population are already actively participating in the

An Exploration into China's Economic Development and Electricity Demand by the Year 2050.
DOI: http://dx.doi.org/10.1016/B978-0-12-420159-0.00007-2

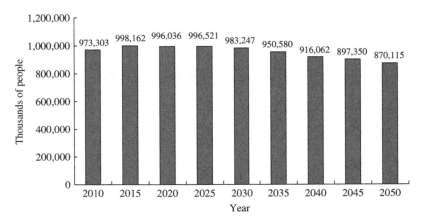

Figure 7.1 China's 2010–2050 working-age population as predicted by the United Nations (program).
Source: Population Division of the Department of Economic and Social Affairs of the United Nations Secretariat, World Population Prospects: The 2008 Revision Population database, http://esa.un.org/unpp/.

labor market activities—so a number of domestic scholars combine these numbers with labor force participation rate trends in order to forecast the future labor supply in China. The 2010–2050 annual labor supply and demand forecasts will segment the working-age population into 15~24, 25~44, 45~64 year-old groups. The general experience of developed countries will then be combined with these numbers and considered with our specific national conditions. This will allow us to estimate the 2050 labor force participation rate for each group. To learn about these forecasts for the future labor supply (Table 7.1).

By 2010, China's total employment is expected to have reached 784 million. This figure will be used as a base, and the adopted future growth rate will be as given in Table 7.1. Therefore, the average labor supply growth rate in the 2010–2050 year period is projected at −0.51%. The labor supply in China by 2040 is expected to be near 65,450 people, and drop to 63,795 people by 2050. Figure 7.2 shows the expected future labor supply and the population.

7.1.2   Fixed Capital Formation Proportional to the Previous Year's GDP

The proportion of gross fixed capital formation in the previous year's GDP has a strong correlation with the investment rate. Changes in the two trends are very similar, as shown in Figure 7.3. For the 1978–2009 year period (the period of China's fixed capital formation), the proportion of gross fixed capital formation of the previous year's GDP averaged about 38.1%, while the average investment rate was about 37.9%. We are assuming that, by 2050, the proportion of the gross fixed capital formation of the previous year's GDP will be higher than the 0.5% investment rate. The future changing trend of the rate of investment will be described later, in detail.

Table 7.1 Forecast of China's Labor Supply (2010–2050)

Year	15~24 Years Old	25~44 Years Old	45~64 Years Old	15~64 Years Old
2010	11,287	37,672	25,600	74,559
2020	7010	35,670	29,568	72,248
2030	7380	34,341	26,615	68,335
2040	6728	29,895	25,621	62,244
2050	5205	30,761	24,704	60,670

Predicted results in the table correspond to the total fertility rate of 1.8.
1. Zhang Yan Qun, Lou Feng. China's economy and long-term growth analysis and forecast of the potential-2020, the number of Technical Economics. 2009 (12):137–145.
2. Wen Wang. Stage of labor supply and demand of low fertility and population change and economic growth in China. China Population Science 2007 (1):44–53.
3. Qi Pearl of 2010–2050 years of labor supply and demand forecasting. Population Research 2010;34(5):76–87.

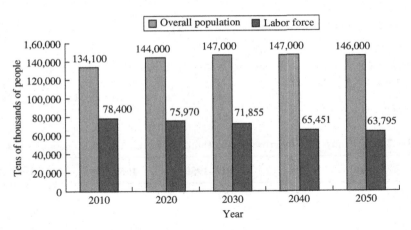

Figure 7.2 China's population in 2050 and labor supply projections.

7.1.3 The Growth Rate of Total Factor Productivity

Empirical studies have shown that total factor productivity—in the economic take-off stage of a country (region)—has experienced a more rapid growth of TFP, its value is 2~3%. Where China is concerned, some scholars have used different methods, and they suspect that the aforementioned 3% average annual growth rate of TFP in China is an overestimation.[4,5,6] With both economic and social development, the TFP growth rate has experienced a gradual downward trend. While the majority of TFP growth in developed countries (after the completion of industrialization) tends to drop below 2%. It has even been less than 1% in some countries (such as the United States, Britain; Table 7.2)—it is assumed that by 2050, China's TFP growth rate will be maintained at about 1.6%.

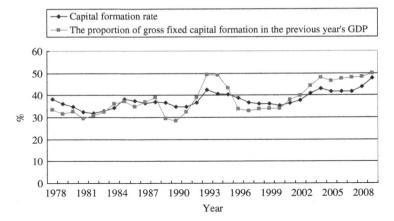

Figure 7.3 Proportion of gross fixed capital formation in the previous year's GDP and investment rate.

1. The population data refers to the National Population and Family Planning Commission of the population development scenarios which set a 2040, 2050 population of 147,000 million and 146,000 million, respectively.
2. Li Jingwen. Productivity and economic growth in America and Japan. Beijing: China Social Sciences Press; 1993.
3. Young A. The lessons form the East Asian NICs: a contrarian view. European Economic Review 1994;2(38):964–973.

Table 7.2 Economy and TFP growth in the United States, the United Kingdom, and Japan

Year		1913–1950	1950–1973	1973–2003
United States	GDP	2.84	3.93	2.93
	TFP	1.62	1.75	0.91
United Kingdom	GDP	1.19	2.93	2.15
	TFP	0.81	1.48	0.65
Japan	GDP	2.21	9.29	2.62
	TFP	0.20	5.12	0.63

Source: Li Shantong. China's economy in 2030. Beijing: Economic Science Press; 2010.

7.1.4   The Proportion of the Three Major Demands in 2050

Although the rates of change in national consumption for different income levels are not the same, the overall trend reveals that the consumption rate with the economic growth by low income to high income shows a U-shaped trend. The consumption rate rises and falls progressively, and countries with different income levels are at different stages of this U-shaped trend.[7] Low-income economies are located at the U-shaped front end, although the rate of consumption in low-income economies gradually decreases, the consumption rate of these countries remains at a high level. From 1965 to 2005, the consumption rate stabilized at around 80%. Furthermore,

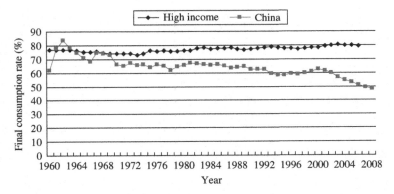

Figure 7.4 The comparison of Chinese consumption to the world's high-income countries. *Source*: Data from WDI2009.

the investment rate hovered in the upper to lower 20% for a long time. Only when the level of economic development of low-income countries reaches the stage of middle and lower income levels will their higher consumption rates decrease significantly. Middle-income economies exist in the middle part of the U-shaped trend. Consumption first decreases, and then increases, and the trend of investment rises and then drops. The consumption rate was near 80% during the late 1960s and gradually decreased as we approached the late 1980s (72~74%). Gradually, it began to increase to a higher level, reaching the current 83%. The investment rate during the 1960s was about 20% and gradually rose to 25% by the early 1980s. It then declined slightly and stabilized at around 20%. High-income economies reach the top rear of the U-shaped trend. There is a steady increase in the rate of consumption, while the rate of investment trend changes from flat to negative. Among these countries, the consumption rate in the late 1960s (72~74%) has steadily climbed to the current level of 80%. The rate of investment during the late 1960s rose from 25% to 27% and then steadily declined to the current approximate level of 20%. As shown in Figures 7.4 and 7.5, since 1960, China's investment rate has shown an overall upward trend, and the consumption rate has shown an overall downward trend. Also, when compared to the high-income countries of the world, China's investment rate has been higher than the average, and the consumption rate has been significantly lower than the average. At present, the changes in China's consumption rate are in the intermediate stage of the "U" shaped law. Therefore, the consumption rate will rebound, while the investment rate will gradually decline, and the gap between China and the world's high-income countries will be gradually reduced.

Figure 7.6 shows the relationship between the countries' export dependency and their total GDP. The figure shows that export dependence in larger economies is generally low, mostly less than 50%—for example, in 2006, US export dependency was 11.2%, Japan's 16.1%, and Germany's 46.7%. Britain, France, Italy, Spain, all demonstrated about 25% dependence. Thus, when compared with the economies of scale, China's dependence on foreign trade has a lower potential for

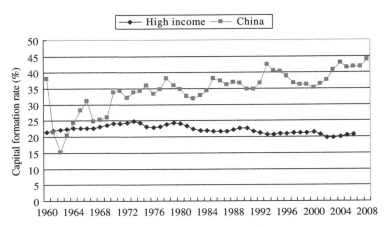

Figure 7.5 The comparison of Chinese investment to the world's high-income countries.
Source: Data from WDI2009.

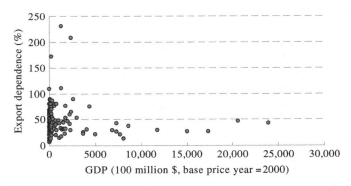

Figure 7.6 The relationships between some countries' export dependency and their GDP in 2007.
Source: Data from WDI2009.

further increases. In 2050, China will have entered the ranks of developed countries. According to the above analysis, when China's investment rate has dropped significantly, and the consumption rate has increased significantly, dependence on foreign trade will have improved significantly. Estimates show that by 2050, China's consumption rate will be 74.5%. The investment rate will have fallen to 24.2%, while the export and import shares of the GDP will be 10.8% and − 9.5%. The changes in China's 2050 GDP ratio for three demands are shown in Figure 7.7.

7.1.5   The Direct Consumption Coefficient of Various Industries in 2050

In order to make the latest input–output tables of 2007 closer to the current economic situation, it is necessary to update them and calculate the direct consumption

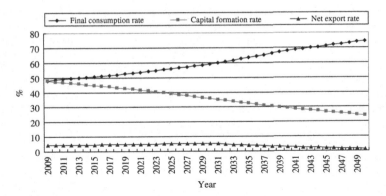

Figure 7.7 The proportion of three major demands to China's GDP in 2050.

coefficient matrix. However, due to the sheer amount of data in the direct consumption coefficient matrix (such as here with 15 industry input–output tables), there are $15 \times 15 = 225$ numbers of its direct consumption coefficient matrix needed to additionally be determined. These rows and columns need to meet certain constraints. Moreover, it is difficult to individually determine the changes in the magnitude in these coefficients over the next 40 years.

By comparing the input–output tables of the United States, Japan, and other developed countries throughout the years, it can be easily seen that the changing of economy structure tended to slow down and the direct consumption coefficient among different industries tended to be more stable. From the preceding analysis, it is not difficult to see that China's current industrial structure is lagging behind that of the United States by at least 50 years. Assuming that the direct consumption coefficient between various industries in China will have gradually stabilized after 2030, and the correlations of input–output among different industries become similar to the current situation of the United States, therefore we are able to use the direct consumption coefficient of the input–output table of the United States in 2007 to adjust the current input–output coefficient of China to finally obtain the direct consumption coefficient of China in 2050.

7.2 Economic Growth by 2050

Table 7.3 gives China's forecasted 2050 GDP results. In 2040, China's GDP will reach about 182 trillion yuan (in 2005 Yuan). During 2030–2040, the average GDP growth will be about 4.8%, which is calculated in accordance with the 2005 exchange rate (100 USD to 819.17 Yuan). In 2040, China's GDP will approach 22.2 trillion USD, and the GDP per capita will be around 15,100 USD. By 2050, China's GDP will reach about 2.73 million billion yuan (33.3 trillion USD), and GDP per capita will be about 187,100 yuan (about 22,800 USD). The average 2040–2050 GDP growth rate will reach 4.2%.

Table 7.3 China's 2050 GDP Forecast Results (in 2005 Prices)

Target	2010	2020	2030	2040	2050
GDP (100 million Yuan)	314,143	672,126	1,137,298	1,817,676	2,731,806
GDP per capita (10,000 Yuan)	2.34	4.67	7.74	12.37	18.71
GDP (100 million USD)	38,349	82,050	138,835	221,892	333,485
GDP per capita (10,000 USD)	0.29	0.57	0.94	1.51	2.28

The exchange rate is in accordance with the 2005 exchange rate, namely 100 USD to 819.17 Yuan.
Prior to 2030 data from the Chapter 5 "scenario" results.

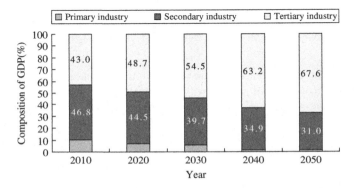

Figure 7.8 Proportion of Chinese GDP by 2050.

It is important to note the changes in industrial structure (as shown in Figure 7.8) in 2040. The proportion of Chinese three industries is 1.9:34.9:63.2. Subsequently, the proportion of primary industry change is very small, in proportion to the secondary industry, and it continues to decline. The tertiary industry's share continues to rise all the way through 2050. At this time, the economic structure will reach a ratio of 1.4:31:67.6. During the 2030–2050, the secondary industry falls 4.4%, which is 8.7 percentage points lower than the primary industry. The tertiary industry rises by 13.1 percentage points during this time.

In the subindustry structure (Table 7.4), in 2050, the industrial value-added share of GDP is projected at 24.7%. Here, the extractive industries will account for 2.2%; food and tobacco industry 2.1%; the textile industry 1.5%; the power industry 2.1%; the chemical industry 2.7%; nonmetallic mineral products industry 0.7%; metal smelting and products industry 2.7%; and the equipment manufacturing industry 7.4%. The building value-added share of GDP was 6.5%; transportation, post, and telecommunications commercial catering value-added share of GDP is projected at about 10%. Finance, real estate, education, and others' value-added share of GDP was 47.4%. When compared with 2030, industry structure, the proportion of the extractive industries, textiles, nonmetallic mineral products, and metal smelting industry, the equipment manufacturing industry declined more significantly.

Table 7.4 China's 2040, 2050 Subsector Economy Structure (%)

Industry	2040	2050	Industry	2040	2050
Agriculture	1.9	1.4	Metal smelting and product manufacturing	3.2	2.7
Mining	2.4	2.2	Equipment manufacturing industry	8.4	7.4
Food and tobacco	2.6	2.1	Other industries	2.5	2.0
The textile	1.8	1.5	Construction	7.6	6.5
Electric power	2.2	2.1	Transport/postal	8.5	9.9
Coke and petroleum processing	0.4	1.3	Commerce and catering	9.8	10.3
Chemical	2.9	2.7	Other Services	44.9	47.4
Nonmetallic mineral products	0.9	0.7			

Agriculture includes agriculture, forestry, animal husbandry, and the fishery. Extractive industries include the coal mining and washing industry, oil and gas exploration, ferrous and nonferrous metals mining and dressing, as well as other mining and dressing industries. The food tobacco industry includes the food manufacturing and tobacco processing industry. The textile industry includes textiles, clothing, shoes of leather, down and other related industry products. The power industry includes electricity, and the heat production and supply industry. The metal smelting and product manufacturing industry includes the metal smelting and rolling processing industry, as well as fabricated metal products. The equipment manufacturing industry includes general/special equipment manufacturing, transportation equipment manufacturing, electrical machinery and equipment manufacturing, communications equipment, computers and other electronic equipment, instrumentation and cultural office machinery. Other industries include handicrafts and other manufacturing, wood processing and furniture manufacturing, paper making, printing and stationery, sporting goods manufacturing, the scrap waste industry, gas production and supply, as well as the water production and supply industry. The transport postal services industry includes transportation and warehousing and postal services. The commercial catering industry includes wholesale and retail trade, and the accommodations and catering industry. Other services include banking, real estate, information transmission, the computer services and software industry, leasing and business services, the research and experimental development industry, comprehensive technical services, water, environment, and public facilities management, resident services and other services, education, health, social security and social welfare, culture, sports and entertainment, public administration and social organizations.

They showed declines of 3.2, 1.7, 2.3, 0.7, and 2.1 percentage points. Conversely, the transport, postal, financial, and other services increased significantly.

GDP growth is an important indicator that can be used to measure a country's economic growth. There are certain links between GDP growth and GDP per capita. Figure 7.9 is a scatter plot which represents the per decade average GDP growth rate since 1960 for major developed countries. It includes the levels of GDP per capita data. The implications which can be seen are that, in the future, most countries show industrialized mid-term GDP growth rates that drop significantly. When GDP per capita reaches 10,000 USD (in constant 2000 $), then the GDP growth rate from the previous decade is more than 5%. When GDP per capita reached 10,000 USD, the GDP growth rate from the previous decade is generally about 4%. With an increase in GDP per capita, the GDP growth rate continues to decline and eventually stabilizes.

Figure 7.10 shows GDP growth in the United Kingdom, the United States, Japan, and South Korea since 1960. While the United Kingdom and the United States are the representatives of the mature industrialized countries, Japan and Korea are the representatives of the newly industrialized countries. In general, in more developed countries, the GDP growth rate (over the same period) is high in comparison to other countries. This is true even during complete industrialization, such as the

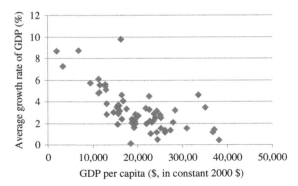

Figure 7.9 The relationship between the GDP growth and the GDP per capita in some developed countries. Countries include Australia, Austria, Belgium, Canada, France, Germany, Italy, the Netherlands, Sweden, the United Kingdom, the United States, Japan, South Korea. Countries' GDP growth in the year 1960–1970, 1970–1980, 1980–1990, 1990–2000, 2000–2009 are average growth rates. Countries' GDP per capita noted for 1970, 1980, 1990, 2000, and 2009.
Source: Data from WDI2010.

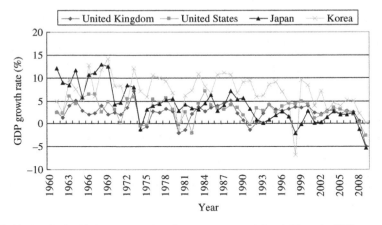

Figure 7.10 The United Kingdom, United States, Japan, and South Korea's GDP growth rate since 1960.

2000–2009. Here, the average annual GDP growth rate of the mature countries was 1.39%, while South Korea's average annual GDP growth rate was 3.51%.

China's 2040 and 2050 GDP per capita are projected to be 15,100 and 22,800 USD (in constant 2005 $), respectively. This corresponds to the average growth rate of 4.8% and 4.2% every decade. From the above analysis, we can still assume that the trend of China's GDP growth rate will be in line with the general laws affecting most developed countries.

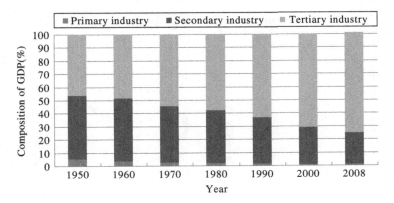

Figure 7.11 Changes in the United Kingdom's economic structure.

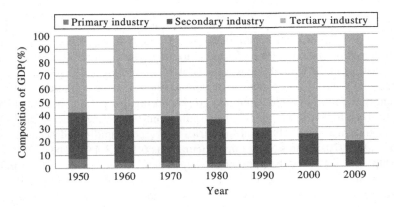

Figure 7.12 Changes in the United States' economic structure.

For the changes in economic structure, the proportions of primary industry and secondary industry of the developed countries are low; the service sector proportions would be higher and further enhancing with economic development.

Figure 7.11 shows the changes in economic structure in the United Kingdom from 1950 to 2008. The proportion of three industries in the United Kingdom in 1950 was 5.7:48.0:46.3 and in 1980 it became 2.2:40.2:57.6. During the 1980–2008, the United Kingdom's economic structure changed relatively quickly. This was especially true for the secondary industry where the proportion experienced a greater decline. Here, the proportion of first industry decreased by 1.4 percentage points; the proportion of secondary industry dropped by 17.6%; and the proportion of tertiary industry rose by 19.0 percentage points.[8] In 2008, the ratio of the three industries was 0.8:22.6: 76.6.

Figure 7.12 shows the three changes in economic structure for the United States from 1950 to 2009. In 1950, the primary industry's proportion of national GDP was

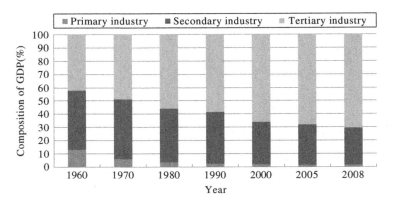

Figure 7.13 Changes in the Japanese economic structure.

6.8%; secondary industry accounted for 35.6%; and tertiary industry accounted for 57.6%. Since the 1980s, the economic center of gravity in the United States has significantly sped up the rate of transfer to the tertiary industry. In 2009, the ratio of the three industries in the United States was about 0.9:18.6:80.5.

Figure 7.13 shows the change in Japan's economic structure since 1960. Japan experienced a period of rapid economic growth during this period, and the economic structure was focused on heavy industrialization. Steel, machinery, petrochemical, and other heavy and chemical industries developed rapidly and became leading industries. In 1970, Japan's economic structure was 6.1:45.3:48.6. The 1970's oil crisis prompted Japan to make appropriate adjustments in economic policy and economic structure. In 2000, Japan's economic structure has changed to 1.8:32.4:65.8. When compared to 1970, primary industry and secondary industry have decreased, while the proportion of the tertiary industry has increased substantially. With the development of high-tech and industrial restructuring in the new century, Japan's proportion of secondary industry continued to fall, and the proportion of the tertiary industry was raised. In 2008, the ratio of the three industries in Japan was 1.5:27.9:70.6.

Figure 7.14 shows the changes in the economic structure of South Korea since 1965. After World War II, South Korea was still a backward agricultural country and gradually transformed into a more developed modern economy. The economic structure here has undergone significant changes. In 1965, primary industry, secondary industry, the tertiary industry had a ratio of 39.4:21.3:39.3. By 2009, the economic structure had reached a ratio of 2.6:36.5:60.9.

These figures show the changes in economic structure of the United Kingdom, the United States, Japan, and South Korea. From the figures, we can see that the changing trends of Chinese economy structure will become closer to those of developed countries in 2030 and will be very similar to the current Japanese economy structure by 2050, but the proportion of tertiary industry will still be lower than those of the United Kingdom and United States.

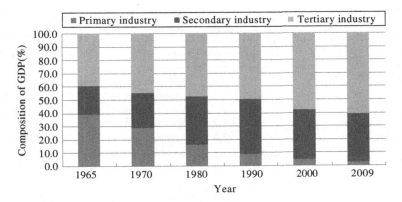

Figure 7.14 Changes in the South Korean economic structure.

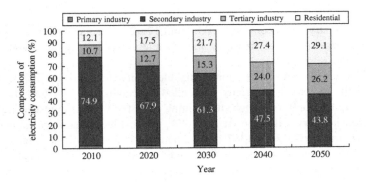

Figure 7.15 The proportion of Chinese electricity demand by 2050.

7.3 Electricity Demand by 2050

In 2040, China's electricity demand will reach 12.1 trillion kWh. At this time, the electricity demand per capita and residential electricity demand per capita will reach 8230 and 2255 kWh, respectively. During the 2030–2040, the average growth rate of electricity demand for the entire country will be 2.12%. Electricity demand elasticity will be about 0.44 with electricity intensity decreased by about 23%. In 2050, China's electricity demand will reach 14.3 trillion kWh, when the electricity demand per capita and residential electricity demand per capita will reach 9813 and 2851 kWh, respectively. During the 2040–2050, the entire country's average growth rate for electricity demand will be 1.7%. The electricity elasticity will be 0.41 and the electricity intensity will have decreased by about 21%.

From Figure 7.15, we can see that the proportion of electricity demand in primary industry, secondary industry, tertiary industry, and residential will be 1.1%, 47.5%, 24.0%, and 27.4%, respectively in 2040; and 0.9%, 43.8%, 26.2%, and 29.1%, respectively in 2050. During the 2030–2050, the proportion of primary and

Table 7.5 Forecast of Proportion of Chinese Subindustries' Electricity Demand in 2040 and 2050 (%)

Industry	2040	2050	Industry	2040	2050
Agriculture	1.1	0.9	Metal smelting and product manufacturing	12.1	10.5
Mining	2.4	2.2	Equipment manufacturing	5.2	4.7
Food and tobacco	1.4	1.2	Other industries	2.3	1.9
Textile	2.3	2.0	Construction	1.4	1.2
Electric power	10.0	9.7	Transport and postal	3.8	4.6
Coke and petroleum processing	0.4	1.3	Commerce and catering trade	5.1	5.5
Chemical	7.8	7.4	Other services	15.1	16.2
Nonmetallic mineral products	2.3	1.8	Residential	27.4	29.1

The range of industries in Table 7.4 is the same.

secondary industry will decrease by 0.88 and 17.42 percentage points, respectively, but the proportion of tertiary industry and residential will rise by 10.95 and 7.35 percentage points, respectively.

Based on the proportion of subindustry electricity demand (Table 7.5), the proportion of the industrial electricity demand will account for 40.4% of the national total demand by 2050, of which, the mining industry will account for 2.2%, food and tobacco industry will account for 1.2%, the textile industry and power industry will account for 2.0% and 9.7%; the chemical industry will account for 7.4 %; the nonmetallic mineral products industry will account for 1.8%; metal smelting and products industry will account for 10.5%; the equipment manufacturing industry will account for 4.7%; the construction of electricity will account for 1.2%; transport and telecommunications and the commercial catering electricity will account for 4.6% and 5.5%, respectively; finance, real estate, education, and other services of electricity will account for 16.2%.

Similar trends can be found in GDP growth. There are certain relationships between electricity consumption growth rate changes and changes in electricity consumption per capita. Figure 7.16 is drawn by the scatter plots of the growth rate (10 years) of national electricity consumption and average electricity consumption per capita in major developed countries. From the figure it can be seen that in most countries, the electricity consumption growth rate dropped significantly after entering the industrialized midterm. Higher electricity consumption growth rate with per capita consumption reached 5000 kWh, generally more than 6%. When the electricity consumption per capita was in the 5000~10,000 kWh scope, the electricity consumption growth rate is generally low, generally less than 3%. The electricity consumption growth rate will decline further when the electricity consumption per capita in the range of 10,000~20,000 kWh. Of course, some countries with high electricity consumption per capita have their own particularity. One such example is Canada in 1970. Here, electricity consumption per capita reached 9832 kWh, and in 1980 it reached 15182 kWh. However, electricity consumption in Canada changed

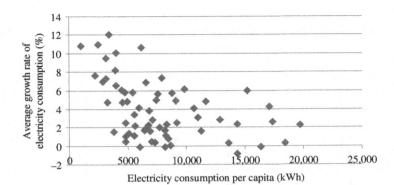

Figure 7.16 The major developed country's electricity consumption growth rate and electricity consumption per capita. Countries include Australia, Austria, Belgium, Canada, France, Germany, Italy, the Netherlands, Sweden, the United Kingdom, the United States, Japan, and South Korea. National electricity consumption growth rates between 1960 and 1970, 1970 and 1980, 1980 and 1990, 1990 and 2000, 2000 and 2009 yearly average growth rate. Countries in electricity consumption per capita for 1970, 1980, 1990, 2000, 2009 electricity consumption per capita.
Source: WDI2010, the IEA.

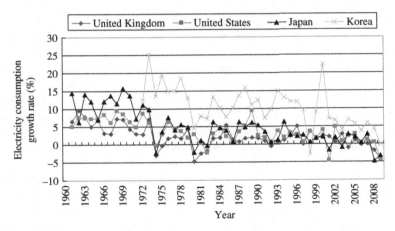

Figure 7.17 The electricity consumption growth rate in the United Kingdom, the United States, Japan, and South Korea since 1960.

during the 1970s and the average annual growth rate was 5.94%. Furthermore, electricity consumption per capita in Sweden reached 11636 kWh in 1980 and 15815 kWh in 1990, but the electricity consumption of Sweden in the period from 1980 to 1990 experienced an average annual growth rate of 4.24%.

Figure 7.17 shows the electricity consumption growth rate of the United Kingdom, the United States, Japan, and South Korea since 1960. It can be seen that in South Korea (as a later developed country), the electricity consumption growth rate is significantly higher than the other three countries. For the United Kingdom, the United States, and Japan after the 1970s (especially after the 1980s), the electricity consumption growth rate slowed down.

From 1973 to the beginning of the early 1980s, the United Kingdom's economic development faced times of crisis, and electricity consumption growth was slowed down. In some years, there was even negative growth in electricity consumption. However, during the middle of the 1980s, the economy maintained a steady low development, and electricity consumption also slowly grew. After entering the twenty-first century, the use of electricity slowed further. During 1980–2000, the United Kingdom's electricity consumption annual growth rate was 1.6%. In 2000–2008, electricity consumption's annual growth rate was 0.4%.

The United States experienced twice oil crises in the 1970s. Due to this, they became aware of the importance of energy and resource conservation. Due to resource efficiency and energy efficiency, labor productivity has increased, but the electricity consumption growth has gradually slowed down. From 1970 to 1990, the annual electricity consumption growth rate was 3.7%. During the 1990s, the United States led information and globalization, and high-value added, low-electricity intensity industries have become the leading industries for promoting the economic development. Here, the growth rate of electricity consumption continued to slow down. During 1990–2009, the average annual electricity growth rate was 1.5%.

Japan's electricity consumption in the late 1960s and after the WWII showed rapid growth. After Japan entered the 1970s, through the oil crisis, the economic development method changed, and the demand for electricity also decreased. From 1973 to 1995, the annual average growth rate was 3.4%. After 1995, Japan's economic development continued to slump, with the electricity consumption growth rate slowing down further. From 1995 to 2009, the electricity growth rate was 1.3%.

When South Korea was experiencing a rapid economy growth during the 1960s and 1970s, this was a further acceleration in the growth in electricity consumption. At this time, there was an average annual growth in electricity consumption by about 16%. In the 1980–1997, economic development entered a period of rectification and coordination of development, industrial restructuring, and an economic slowdown. This made the electricity growth rate down significantly, with an average annual growth of 11.6%. During 1997–2008, the tertiary industry experienced rapid development, especially in consumption on behalf of the service sector which further accelerated the development. During the impact of the Asian financial crisis, the electricity growth rate fell further, with an average annual growth rate of only 6.95%.

During the 2030–2050 year period, the average growth rate of China's electricity demand will be around 1.91%. This rate is lower than the average growth rate of electricity demand in the United States and Japan from 1985 to 2005, but higher than average growth of demand in the United States and Japan from 1990 to 2009.

From the point of view of the structural changes in the electricity consumption of three industries and residential, the overall trends of the change in power structure of developed countries are as follows. First, the electricity proportion of primary industry has been declining, the electricity consumption of secondary industry first

Table 7.6 Proportion of Electricity Consumption in the United Kingdom, the United States, Japan, and Korea (%)

Country	Industry	1970	1980	1990	2000	2007
United Kingdom	Primary	1.69	1.16	1.35	1.28	1.08
	Secondary	43.35	39.49	38.89	36.73	36.05
	Tertiary	18.88	24.04	26.78	29.12	30.06
	Residential	36.08	35.31	32.98	32.87	32.81
United States	Primary	–	–	–	–	–
	Secondary	41.00	38.90	34.86	34.33	30.20
	Tertiary	25.30	26.70	31.08	32.45	34.29
	Residential	33.50	34.30	34.06	33.22	35.51
Japan	Primary	–	–	0.41	0.17	0.09
	Secondary	57.55	50.13	50.39	42.98	34.39
	Tertiary	10.12	16.56	24.34	29.90	37.10
	Residential	16.17	17.56	24.86	26.95	28.42
Korea	Primary	0.78	0.61	1.55	2.27	1.99
	Secondary	70.31	70.67	62.28	55.23	51.44
	Tertiary	17.19	12.53	17.37	29.18	32.77
	Residential	11.72	16.19	18.80	13.32	13.80

Source: International Energy and Electric Power Statistics Manual (2009 Edition).

rises and then drops, and the proportion of commercial electricity consumption has been rising. The proportion of residential electricity consumption has also kept rising but with fluctuations. Here, in the intermediate period of industrialization, the rapid industrial development causes a substantial increase in industrial consumption. This resulted in the rise of the proportion of secondary industry. At this stage, the proportion of industrial electricity is significantly higher than that of commercial and residential electricity consumption proportion. After entering later stages of industrialization, the rate of industrial development is reduced, and the development of the tertiary industry is accelerated. At this time there is a substantial improvement in people's living standards, thus the proportion of electricity consumption for the second industry fell, and the proportion of the commercial and residential electricity consumption proportion rose.

Table 7.6 lists proportional changes of the electricity consumption in the United Kingdom, United States, Japan, and South Korea since 1970. It can be seen that for these countries in 2007, the proportion of primary industry's electricity consumption is very low. The secondary industry, tertiary industry, and residential electricity proportion are in the range of 28~38%. It is believed broadly that in these countries, the second and tertiary industries, as well as residential electricity consumption are the "3 biggest parts" of the structure of electricity consumption. In 2007, the proportion of tertiary industry of electricity consumption in South Korea was almost equivalent to above three developed countries. However, the proportion of electricity consumption of secondary industry was significantly higher, while the proportion of residential electricity consumption was significantly lower.

According to the above analysis, China's electricity consumption structure in 2050 would be similar to Japan's in 2000. The secondary industry's electricity consumption would be in the range of 40~50%, and the proportion of the tertiary industry and residential electricity consumption would be in the range of 20~30%.

Notes

[1] Chow. China's economic transformation. Beijing: China Renmin University Press; 2005.
[2] Wang Xiaolu and Fan Gang. The sustainability of China's economic growth—the cross-century review and prospects. Beijing: Economic Science Press; 2000.
[3] Zhang Jun, Zhang Yuan. Re-estimated the Chinese capital stock K of economic research. 2003 (7):35–44.
[4] Zhou Jianjun. China's tax reform recursive dynamic CGE model [doctoral thesis]. Wuhan: Huazhong University of Science and Technology; 2004.
[5] Zhai Fan, Li Shantong, Feng Shan. Medium-term economic growth and structural change—recursive dynamic general equilibrium analysis of the theory and practice of systems engineering. 1999 (2):88–95.
[6] FISCAL Jiajun snow, total factor productivity estimation: 1979–2004. Economic Research 2005 (6):13–19.
[7] Zhao Ping. From the global perspective trend of China's consumption rate. China Business Guide 2010:17–18.
[8] Chen Zhao, Liu Wei. A chronic trade deficit, economic growth and economic cyclical fluctuations—A 1815–1993 analysis of the UK's economic history and international trade. 2009 (2):121–128.

Appendix A: Parameter Settings of the Three Scenarios

For the values of key parameters in each scenario of economic development, see Table A.1 for specific settings.

Table A.1 Scenario Parameter Settings

Scenario Category	Scenario
Common set of scenarios	1. Population trends for the total exogenous, the direct use of the projections of the United Nations. 2. Exogenous growth of the total workforce. 3. Between 2010 and 2030 balance of payments will be gradually adjusted. 4. A variety of domestic tax rates remain unchanged, a variety of transfer payments to the exogenous accounts are fixed (share of government revenue, or the proportion of total corporate income remains unchanged).
	Set as follows:
Scenario 1	1. Urbanization rate—the level of urbanization and urban and rural population, exogenous from 2010 to 2020, experiences an average increase of 1 percentage point. For 2021–2030, the urbanization rate increased by 0.7 percentage points annually. 2. The proportion of the government consumption share of government revenue remains unchanged. 3. Assuming that the 2010–2020 growth rate of total factor productivity remains on par with past, namely to maintain the overall level of about 2%[a]. 4. The rate of change of the preferences of technological progress and intermediate exogenous inputs. 5. World economic growth will maintain the average levels over the past 10 years, or about 3%.

(Continued)

Table A.1 Scenario Parameter Settings

	Set as follows:
Scenario 2	1. To increase government investment in education, medicine, scientific research, and social welfare, and to adjust the structure of the Government's public spending, – i.e., to increase the proportion of the expenditures for education, medicine, scientific research, and social welfare. 2. Accelerate the urbanization process, the gradual elimination of barriers to labor transfers. The urbanization rate between 2010 and 2030 scenarios shows an annual increase of 0.1–0.2 percentage points, to speed up the transfer of urban and rural labor. 3. Increase public spending; government to pay for poor areas and poor transfers. For 2010 to 2030, the scenario is 10–15%. 4. Improve services and regulatory reforms to reduce the service industry tax burden. Between 2010 and 2030 the service sector TFP scenarios show a high of 0.9 %, and gradually make the services sector's tax burden reduced by 10%. 5. World economic growth rate stay at the averages of the past decade, i.e., about 3%
Scenario 3	Set as follows: 1. Slow process of urbanization and slow labor transfer: the urbanization rate from 2010 to 2030 is 0.1 to 0.15 percentage points lower than the same in Scenario 1 each year, and the speed of labor transfer of urban and rural labor force is also slower than that in Scenario 1. 2. As the world economy is slowly recovering, trade protectionism is becoming more serious. Distinct to the short recovery of growth of exporting demand in Scenario 1, China's export demand from international market will recover after "Twelfth Five Years". 2015–2030, subject to the enhancement of trade protectionism, export growth speed lower than Scenario 1. 3. World economic growth was slightly lower than the **Scenario 1** with 2.7% per annum. 4. Technological innovation and efficiency improvements slowed, TFP was about 0.4 percentage points lower than the **scenario I**.

[a]TFP scenarios manufacturing more than the service sector by 0.5–1 percentage point.

In addition to the economic development parameters, when using the LEAP model for scenario analysis of energy and electricity demand, we also need to set certain technical parameters, including product outputs, energy intensity, electricity intensity, etc. The main product output settings are partly shown in Table A.2.

Table A.2 Major Industry Production or Service Volume

Scenario	Name	Unit	2010	2020	2030
Scenario 1	Cloth	100 million m	923	1778	2734
	Crude steel	10,000 t	63,000	90,500	85,500
	Aluminum	10,000 t	1550	2975	2700
	Cement	100 million t	17.9	29.1	34.2
	Fertilizer	10,000 t	6768	9958	11,413
	Paper/ paperboard	10,000 t	9682	20,191	29,702
	Construction area	100 million m^2	64.7	136	204
	Converted turnover	1 trillion tkm	13.6	35.1	55.9
Scenario 2	Cloth	100 million m	923	1583	2258
	Crude steel	10,000 t	63,000	85,900	77,000
	Aluminum	10,000 t	1550	2826	2550
	Cement	100 million t	17.9	27.1	30.9
	Fertilizer	10,000 t	6768	8652	8485
	Paper/ paperboard	10,000 t	9682	19,125	24,289
	Construction area	100 million m^2	64.7	131	171
	Converted turnover	1 trillion t km	13.6	38.2	64.5
Scenario 3	Cloth	100 million m	900	1680	2520
	Crude steel	10,000 t	63,000	83,000	80,000
	Aluminum	10,000 t	1550	2750	2400
	Cement	100 million t	17.9	26	31.2
	Fertilizer	10,000 t	6770	10,100	12,500
	Paper/ paperboard	10,000 t	9680	19,000	28,500
	Construction area	100 million m^2	64	136	177
	Converted turnover	1 trillion t km	13.6	33	53.7

The main energy consumption and electricity consumption parameters are as follows:

A.1   The Iron and Steel Industry

The steel industry is China's biggest power user, since 2000, energy consumption accounting for the proportion of primary energy consumption is rising. It reached 17.8% in 2008. The proportion of the consumption of coal and coal chemical products has remained above 80%. From an energy consumption level, in 2005 a medium-sized enterprise in Chinese steel showed comparable energy at 732 kgce/t, which is about 1.2 times the advanced international advanced level. In 2006, ferroalloy electricity consumption reached 7594 kWh/t, 2.5 times the international advanced level.[1] China's iron and steel industry shows great energy potential for energy savings and emission reductions. Over recent years, with the phasing out of backward production capacities and by implementing technological innovations and a series of energy saving measures, the comprehensive energy consumption of steel has declined. However, when compared with international standards, there is still a wide gap. In the future, the steel industry will continue to eliminate backward production capacity and production processes, thereby further improving the utility ratio of advanced and efficient technology. By 2020, the current widespread use of energy-saving technologies and production processes will be used in China. The proportion of EAF steel will reach about 20% of overall energy consumption, with energy consumption of per ton of steel attaining the advanced international levels of 2005. The level of electricity consumption will be increased. The steel manufacturing process of new and transformed iron and steel production lines have all basically adopted the advanced world level. Furthermore, with the decommissioning of the old production line, the 2030 energy consumption of the iron and steel industry (as a whole), and the technology levels will be close to the advanced world advanced. Meanwhile, with the cumulative remainder of China's steel production continuously improving, the proportions of EAF steel will increase significantly.

A.2   The Nonferrous Metals Industry

There are a wide range of products produced by the non-ferrous metal industry. Of the 10 kinds of commonly used non-ferrous metals, copper, aluminum, lead, and zinc production accounted for more than 90%. Furthermore, more than 80% of total energy consumption by the industry was taken up by these four metals. Aluminum accounted for most of its energy consumption and the electricity consumption level of the top row. Therefore, in the non-ferrous metals industry, energy demand and electricity demand is mainly aluminum based. This remains true even after taking into account the energy consumption changes of the other metals. In 2009, the aluminum AC consumption was 14171 kWh/t, slightly higher than the advanced international level in 2005, and higher than the 2005 Canada Alma, near 1300 kWh/t, the comprehensive energy consumption of copper smelting 509 kgce/t; in 2005 zinc smelting energy consumption is 1953 kgce/t, higher than the same period in the international advanced level nearly 500 kgce/t. From an overall perspective, the non-ferrous metal industry still has certain room for energy-saving potential. With regard to the production process, the smelting of copper is mainly altered to promote a state-of-the-art flash smelting process. This means the phasing-out and transformation of the blast furnace. The reverberate furnaces and other traditional crafts are also gradually eliminated. Aluminum production has vigorously developed large pre-baked anode cell technology. In 2020, electrolytic aluminum electricity consumption is set to reach the international advanced level of other metals. Energy consumption will fall, and the level of overall energy consumption of aluminum will fall faster than the rate of decline of the electricity consumption level.

A.3   The Building Materials Industry

The main products of the building materials industry are: cement, flat glass, brick, and tile. Among these, the total terminal energy consumption from cement accounted for about 40% of final energy consumption of the entire building materials industry. Its electricity consumption accounted for about 50%. In 2009, the comprehensive energy consumption of cement was 139 kgce/t, which is 1.3 times the same period at the advanced international level and still with large potentials for energy saving. From the production process, the new cement production lines, with state-of-the-art dry process kilns and kiln technology, the energy consumption level of the production process of the old plants is also higher. Due to new renovations carried out 2020, China's overall energy consumption by cement will reach the international advanced level, with the promotion of the waste heat utilization technology. By 2030, there will be a further decline in energy consumption, and it will have reached the advanced international level.

A.4   Paper Industry

In 2008, the comprehensive energy consumption from China's paper and cardboard industry was 1153 kgce/t. This is 1.9 times greater than the advanced international level over the same period. Due to the increased popularization of new cooking technology, warm-up recycling, cogeneration technology, and the improvement of the utilization of waste paper recycling, the energy consumption level of China's paper industry will experience a rapid decline. It should reach the 2005 advanced international level by 2015, and then will experience an even faster decline.

A.5 Chemical Industry

This industry includes different types of chemical products, mainly ammonia, caustic soda, soda ash, ethylene, and fertilizer. In 2005, the overall energy consumption of ammonia was 1263 kgce/t, which 1.3 times the international advanced level over the same period. It had an energy consumption level of 1.15 times the advanced international level of caustic soda, and soda ash's energy consumption level was 1.3 times the advanced international level. With elimination and transformation of backward capacity, and with the introduction of new technologies, the energy consumption level of the chemical industry will reach the 2005 advanced international level by 2015. In 2020, it will reach the advanced international level of 2010, and in 2030 it will be on par with the ranks of the international advanced level.

With per capita income levels increasing, people's living standards and quality of life will continue to improve. The ownership rate of household appliances and the average working time will continue to increase. Since the year 2000, for every 100 urban residents and rural residents, the appliances most commonly used in the household have increased.

With the rise in the level of per capita income, the standard of living and quality of life of the people will see constant improvement, and so will the rise in numbers of ownership of electrical appliances as well as average working hours. From the rising trend in the ownership of electrical appliances among both the urban and rural population since year 2000, as well as in-depth analysis of changes in rates of usage and power usage per appliance, the rates of ownership and usage of various kinds of electrical appliances among both the urban and rural population are reflected respectively in Tables A.3 and A.4.

Conventional air conditioners, refrigerators, washing machines, televisions, water heaters, and other household appliances have a relatively high usage rate among urban residents. Some of these appliances have even reached a point of saturation, so the rate of growth of these appliances will not increase. However, the improvements in living comfort and the power and utilization of household appliances will still continue to increase. For rural residents, the ownership rate for air conditioners, refrigerators, washing machines, and water heaters is low, but, in the future it will show rapid growth.

Table A.3 Ownership of Household Appliances by Urban Residents and Related Parameters

Category	Indicator	2010	2020	2030
Air conditioning	Ownership (units/one hundred)	109	140	180
	Average power (w)	1200	1300	1100
	The annual average time (h/year)	350	540	680
Refrigerator	Ownership rate (units/one hundred)	97	100	120
	Refrigerator efficiency (kWh/d)	0.8	0.8	0.8
Washing machine	Ownership rate (units/one hundred)	97	100	100
	Average electricity consumption (kWh/time)	0.6	0.8	0.6
	Weekly washing times (times)	3	5	8
Television	Ownership rate (units/one hundred)	140	180	220
	Average power (W)	200	320	300
	Viewing time (h/(units d))	2.5	3.5	3.2
Lighting	Penetration of energy-saving lamps (%)	55	100	100
	The number of household lighting lamps (40 W fluorescent lamp standards)	10	14	21
Water heater	Ownership rate (units/one hundred)	81	83	75
	Average power (W)	1500	2000	2000
	Average number of hours (h/d)	0.5	1	1
Electric cooker	Ownership rate (units/one hundred)	100	130	140
	Average power (W)	1500	2000	2500
	The number of hours (h/d)	0.15	0.5	0.5
Other appliances	Capacity (W)	1000	2000	1800
	The number of hours (h / d)	1	1	1.5

Household appliances average power usage to refer to China's energy-saving manual (Wang, QingYi's writing).

Table A.4 Ownership of Household Appliances by Rural Residents and Related Parameters

Category	Index	2010	2020	2030
Air conditioning	Ownership (units/one hundred)	15	45	70
	Average power (w)	1200	1300	1100
	The annual average time (h/year)	360	440	540
Refrigerator	Ownership rate (units/one hundred)	41	75	95
	Refrigerator efficiency (kWh/d)	0.6	0.7	0.65
Washing machine	Ownership rate (units/one hundred)	57	88	98
	Average electricity consumption (kWh/time)	0.8	1	0.6
	Weekly washing times (times)	2	4	6
Television	Ownership rate (units/one hundred)	104	130	180
	Average power (W)	200	270	270
	Viewing time (h/(units d))	4	4	3.2
Lighting	Popularization rate of energy-saving lamps (%)	30	70	100
	The number of household lighting lamps (40 W fluorescent lamp standards)	6	10	18

(Continued)

Table A.4 Ownership of Household Appliances by Rural Residents and Related Parameters

Category	Index	2010	2020	2030
Water heater	Ownership rate (units/one hundred)	31	70	100
	Average power (W)	1500	2000	2000
	Average number of hours (h/d)	1	1	1.5
Electric cooker	Ownership rate (units/one hundred)	30	55	70
	Average power (W)	1500	2500	2500
	The number of hours (h/d)	0.15	0.6	0.5
Other appliances	Capacity (W)	1000	1500	1800
	The number of hours (h/d)	0.3	0.5	1

Note

[1] Wang, QingYi, 2009 China's energy data.

Appendix B: CGE Model Applied in This Book

This model is a Dynamic Recursive Chinese CGE developed by Development Research Center of the State Council. This model is recursively dynamic; it solves a series of static equilibriums to simulate the dynamic characteristics of the economic development of the model simulation time period between 2010 and 2030.

B.1 The Department Set

In the national model, there are 42 production sectors (one for the agricultural sector, 24 industrial sectors, a construction sector, and 16 service sectors). These 42 departments are divided mainly based on the national sector in input–output table, but in the regional model, in order to reduce the model dimension and improve processing speed and accuracy in the case of not affecting the main results of the model, the 42 departments are combined with such principles that, wherever possible, they retain the energy and electricity consumption of large departments, the industrial sector was retained as much as possible, while the service sector merged more. In the services sector, in consideration of large consumption of energy and electricity by the transportation industry, the combination of these sectors is saved. For the specific division of sectors, see Table B.1.

B.2 The Model Equations

The model mainly consists of six modules which are production modules, income and demand modules, the module prices, the international trade module, the balanced closed module, and the dynamic module. The model equation set can be found within the literature.[1]

Table B.1 Sector Model

Sequence Number	42 Departments of the Country	26 Departments
1	Agriculture	Agriculture
2	Coal mining	Coal mining
3	Oil and gas mining industry	Oil and gas mining industry
4	Metal ore mining	Metal ore mining
5	Non-metallic mineral mining	Non-metallic mineral mining
6	Food manufacturing and tobacco processing industry	Food manufacturing and tobacco processing industry
7	The textile industry	Textile and wood processing industry
8	Clothing, shoes, down and related products	
9	Wood processing and furniture manufacturing	
10	Manufacturing of paper printing and stationery	Manufacturing of paper printing and stationery
11	Petroleum processing, coking, and nuclear fuel processing industry	Petroleum processing, coking, and nuclear fuel processing industry
12	Chemical industry	Chemical industry
13	Non-metallic mineral products industry	Non-metallic mineral products industry
14	Metal smelting and rolling processing industry	Metal smelting and rolling processing industry
15	Fabricated metal products	Metal products and general, special equipment and transportation equipment manufacturing industry
16	General, special equipment manufacturing industry	
17	Transportation equipment manufacturing industry	
18	Electrical, machinery, and equipment manufacturing	Electrical, machinery, and equipment manufacturing
19	Communications equipment, computers and other electronic equipment manufacturing	Communications equipment and instrumentation equipment manufacturing
20	Instruments meters cultural and office machinery	
21	Other manufacturing	Other manufacturing
22	Scrap waste	
23	Electricity, heat production, and supply industry	Electricity, heat production and supply industry
24	Gas production and supply industry	Water production and supply industry
25	Water production and supply industry	
26	Building industry	Building industry
27	Transportation and warehousing industry	Transportation and warehousing industry

(*Continued*)

Table B.1 (Continued)

Sequence Number	42 Departments of the Country	26 Departments
28	Postal	Posts and
29	Information transmission, computer services and software industry	telecommunications and information transmission industry
30	Wholesale and retail trade	Wholesale, retail, and
31	Accommodation and catering industry	accommodation and catering industry
32	Real estate	Real estate
33	Education	Education and health care
34	Health, social security, and social welfare	
35	Finance and insurance	Other services
36	Leasing and business services	
37	Sciences research	
38	Integrated technical services	
39	Water, environment, and public facilities management industry	
40	Resident services and other services	
41	Culture, sports, and recreation	
42	Public administration and social organizations	Public administration and social organizations

B.3   The Set of Exogenous Parameters

Population and labor force growth:

For the total population and age structure, labor force changes in the total labor transfer have an important impact on population growth by family planning policies. This also involves improved living standards, lifestyle changes, and many other factors, but it mainly impacts national population policy. In the CGE model, the population growth as the exogenous variables of the model is the choice of the World Bank forecast data on the population and working age. It says that China's population peak will appear in about 2032, and that the working-age population will peak in the 2016–2027 period. The working-age population will be about one billion people, and the labor peak of the total resources of the national workforce will be about 820 million people. This will be an increase of 40 million to the labor force population in 2006. To see China's total population, as well as its working-age population, according to the Chinese Academy of Social Sciences forecast program scenarios, see Figure B.1.

B.4   The Household Savings Rate

Savings has an important impact on economic growth. Savings can be divided into household savings, corporate savings, and government savings. Household savings is

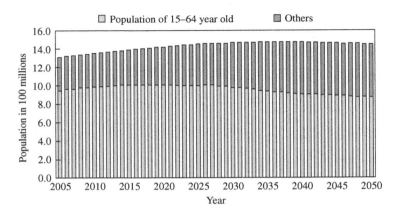

Figure B.1 Chinese Academy of Social Sciences Population forecast program scenarios for China's total population and working age population changes.

one of the most important parts; however, it is also the part that experiences the most changes. For each sector's savings percentage of GDP during the 1992–2007 period, see Table B.2. According to existing studies, many factors exist that affect the development and changes in the household savings rate.

To predict possible changes in China's household savings rate and settings, one should first analyze the main factors which affect changes in the savings rate.

The uncertainty of resident's wealth constraints and future expenditures will affect the residents' savings behavior. Gao Mengtao (2008) made a study of China's eight provinces with rural household income consumer micro-data research[2]. They found that the savings behavior of China's farmers was affected by significant liquidity constraints and the preventative motive. These could both significantly improve the farmers' savings and reduce the level of consumption of the farmers. Liquidity constraints for low-income people improve the effect of the savings in a more obvious manner. Therefore, it can be expected to raise the level of income of rural residents and cause social security system improvements. The level of consumption of rural residents will have to rise further, or the savings rate will face a certain reduction.

The level of social security will also affect the household savings rate. In a sound national social security system, savings are an important supplement in the form of endowment insurance. For example, for the average pension motive in 2000, China's household savings was as high as 21.4%. Therefore, we can expect that with the gradual development of China's social security system, the resident's use of that part of the pension savings will gradually be reduced.

Residents in the age structure changes will also affect the savings rate. In general, in a minor population unable to save, an aging population makes the savings rate lower. As the proportions of aging population increase, this will result in a lower savings rate.

In terms of changes in other countries of the world, the household savings rate will be some a reference set of the household savings rate in China. Figure B.2

Table B.2 China's Different Savings as a Percentage of GDP (%)

Year	Household Savings Rate	Corporate Savings Rate	Government Savings Rate	National Savings Rate	Residents' Savings as a Percentage of Disposable Income Ratio
1992	21.14	13.37	5.90	40.41	31.12
1993	19.32	16.15	6.24	41.71	29.91
1994	21.50	16.03	5.22	42.75	32.57
1995	19.75	16.45	4.81	41.01	30.00
1996	21.05	13.39	5.36	39.81	30.77
1997	20.43	15.93	4.13	40.49	30.46
1998	20.13	14.14	5.20	39.46	29.93
1999	18.30	14.12	5.68	38.10	27.63
2000	16.38	15.53	6.31	38.22	25.45
2001	16.03	15.00	7.50	38.54	25.37
2002	18.60	14.30	7.23	40.14	28.59
2003	18.23	15.58	9.38	43.20	28.89
2004	18.49	22.00	6.08	46.57	31.65
2005	21.50	20.36	6.36	48.22	35.61
2006	21.73	18.83	8.92	49.48	36.40
2007	22.24	18.77	10.84	51.85	37.94

1. 1992–2001 data source: RenRuoen, Qin Xiao, China and the United States than the residents of the savings rate measured at 1998–2009, Freeones-2001, Economic Research, 2006 (3).

2. 2002–2007 data source: 2004–2009, China Statistical Yearbook, Flow of Funds Table (Physical Transaction).

shows a part of some country's household savings rate. The figure shows that the household savings rate in many countries has experienced a stage of increase and then decrease. Relatively speaking, in East Asia, and in South Korea, the household savings rate was higher, and during the twentieth century it reached the level of around 25% in the 1980s and 1990s. When compared with those in developed countries, China's household savings rate was much higher. The above analysis forecasts that China's household savings rate will gradually decline with economic development and social progress. In the model, according to the settings, the household saving rate will drop from current 38% by about 13% which is around 25% by 2030.

B.5 Total Factor Productivity

The total factor productivity is the most important factor for long-term impact on economic growth. Many scholars both at home and abroad differ on China's TFP estimates (Table B.3). This is due to the use of data and differences in estimation

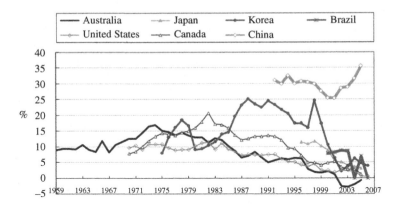

Figure B.2 Part of the national household savings rate changes.
Source: OECD Fact book 2008: Economic, Environmental and Social Statistics.

methods. The results also differ since China's reforms and opening up, and China's average annual TFP growth rate is approximated to be in the 2–4% range.

The future trend of TFP growth, on the one hand, depends on development and changes in the domestic TFP factor. On the other hand, it depends on the changes in China and the world technology gap. From the domestic impact factor, as concluded from the 2009 study of the Wang Xiaolu, we can see many aspects of the factors that have a significant effect on TFP. These include the Human Capital Spillover technology capital, market-oriented reform, urbanization, foreign effect, foreign trade effect, the basis of facilities, administration costs, and the final consumption rate. According to the most likely scenarios based on an extension of the current changes in the

Table B.3 China's TFP Growth Estimates

Guo, QingWang, Jia, JunXue	1979–2004	0.89%
Sun, LumLum, Ren, RuoYen	1981–2002	3.14%
Wang Xiaolu, Fan, Gang, Liu Peng	1999–2007	3.63%
Jefferson and Rawski (1994)	1980–1992	2.4%
Hu and Khan (1997)	1979–1994	3.9%
Wang and Hu (1999)	1977–1995	2.9%
Chow (2002)	1977–1998	2.7%
Heytens and Zebregs (2003)	1990–1998	2.7%
CSLS (2003)	1980–2000	1.7%
Wu (2004)	1982–1997	1.4%
Kuijs and Wang (2006)	1993–2004	2.7%
Hong Kong Monetary Authority (2006)	1977–2003	2.9%
CEM update of Kuijs and Wang[a]	1993–2005	3.0%
Hofman et al.		3.0%
Bosworth and Collins (2007)[b]	1993–2004	4.2%

[a]TFP growth rate of state-owned enterprises.
[b]Use the revised GDP data.

trends of these contributing factors, and the appropriate changes, including consideration of the current world economic crisis, the potential impact is that the TFP average growth may be around 1.79% in 2007–2020. Where there is strengthening of institutional suppression policy management costs, enhanced education and training, improved social security and public service systems, and an inhibition of the widening income gap, average growth speed of TFP is expected to reach the higher level of 3.95% in the period between 2007 and 2020.

As can be seen from the trend of changes in the rate of growth of the TFP in various countries all over the world (Table B.4), when the economy has developed to a certain level, the TFP will exhibit a downward trend, taking for example, Japan, Germany, and the United States, which have all showed a lower TFP; and Korea during the 1990s, in the twentieth century has also displayed the same downward trend, Singapore is the only exception with a rise.

Referencing countries in the world with the TFP development of general laws, and by taking into account China and the countries with cutting edge technology, the technology gap is an important factor for affecting TFP change. The overall view was that China's TFP may be in a long-term downward trend in book research. The 2030 TFP growth rate is set to slightly decrease to about 2.1%.

B.6 The Rate of Change of Intermediate Inputs

For the vast majority of departments in terms of the national economy, intermediate inputs often accounted for more than half of the sector's total output. Therefore, the intermediate inputs arising from changes in the structural change effect is sometimes greater than the initial investment. Furthermore, the role of final consumption and intermediate inputs are an important part of the national economic structure. With economic development and technological progress, the intermediate input rate of each industry will change. From the CGE model of the book, according to China's 1992–2005 input–output table changes, and by referring to the United States and Japan in the middle input rate variation (Table B.5), the CGE model medium- and

Table B.4 General Trends of Growth Rate of National TFP (%)

Year	1951–1960	1961–1970	1971–1980	1981–1990	1991–1995
Thailand	–	2.61	2.13	2.62	2.05
Singapore	–	3.88	2.74	3.35	4.81
Korea	–	3.22	2.46	5.01	3.18
Japan	2.80	2.64	0.58	0.63	−0.13
Germany	4.43	1.53	0.82	0.71	−2.05
The United States	1.52	1.78	0.26	0.51	0.53

Source: Mercer 2 Melbourne Institute, 1997, quoted from Guo, QingWang, Jia, JunXue, 2005.

Table B.5 The Intermediate Input–Output Coefficients in China, Japan, and the United States (%)

Sector	China				Japan				The United States	
	1992	1997	2002	2005	1970	1980	1990	2000	1995	2000
Agriculture	35.7	40.4	41.8	41.5	36.9	46.4	43.0	46.3	60.3	61.5
Coal mining	56.1	48.6	43.1	56.0					52.5	41.1
Oil and gas mining	37.8	26.2	28.9	30.7						
Metal ore mining	60.7	64.6	56.9	66.0	36.4	49.7	48.4	43.3	48.6	48.5
Non-metallic mineral mining	60.6	58.6	53.5	69.6						
Food manufacturing and tobacco processing industry	74.3	72.3	68.9	72.3	73.1	71.8	67.8	76.2	66.9	72.2
The textile industry	79.4	71.8	75.2	79.1	73.9	70.9	66.1	66.9	66.4	65.9
Garment and leather, down and related products manufacturing industry	78.8	68.8	75.4	75.0						
Wood processing and furniture manufacturing industry	74.7	72.1	72.7	76.7	72.7	71.6	65.5	67.1	61.9	66.6
Manufacturing of paper printing and stationery	73.0	68.5	66.3	75.4	66.7	68.4	60.8	62.3	53.5	56.4
Petroleum processing, coking, and nuclear fuel processing industry	72.0	77.9	82.8	81.2	58.7	84.8	62.5	79.9	85.9	88.6
The chemical industry	72.1	73.1	73.1	78.2	69.7	79.7	72.1	76.2	63.9	64.1
Non-metallic mineral products industry	65.3	68.4	67.1	73.2	61.4	68.4	59.6	61.3	77.3	
Metal smelting and rolling processing industry	69.3	79.6	75.6	79.4	80.3	80.0	75.9	76.7	70.1	80.9
Fabricated metal products	76.0	76.7	76.3	78.0	57.9	62.5	57.6	58.6		
General, special equipment manufacturing industry	71.7	66.4	71.9	76.0	64.8	67.0	60.9	68.6	54.5	61.2
Transportation equipment manufacturing industry	73.3	73.8	73.8	78.6	64.6	64.1	66.3	71.6	62.8	

Electrical machinery and equipment manufacturing industry	74.6	77.7	75.9	79.2	66.7	67.6	64.1	68.5	55.3	58.4
Communications equipment, computers and other electronic equipment manufacturing industry	75.0	74.6	79.0	84.3	65.9	66.9	69.1	71.3	51.2	
Instruments meters cultural and office machinery	66.1	68.7	74.3	78.4	46.3	56.4	57.9	65.3	73.7	62.9
Other manufacturing	75.6	65.7	71.9	73.5	66.7	64.3	64.8	75.7		
Waste and scrap	66.7	0.0	0.0	0.0						
Electricity, hot water production and supply	51.2	56.8	49.9	69.1	35.1	57.9	44.9	48.1	44.7	40.7
Gas production and supply industry	83.1	73.7	79.6	74.2					75.8	
Water production and supply industry	51.1	50.0	50.0	55.2					39.5	
Construction	70.4	71.3	76.6	74.6	64.8	60.1	55.8	56.8	53.7	49.4
Transportation and warehousing industry	44.2	44.8	51.6	56.9	40.6	52.9	46.7	43.7	51.7	50.6
Postal	31.2	58.6	60.0	53.8	16.1	21.9	22.4	45.6	39.8	51.1
Information transmission, computer services and software industry		40.7	43.9	52.1						
Wholesale and retail trade	54.9	49.0	45.9	30.0	32.3	33.4	32.9	36.3	34.8	33.2
Accommodation and catering industry	59.8	57.7	59.5	62.6	59.2	54.1	52.1	55.6	51.3	47
Finance and insurance	47.8	39.0	36.1	38.5	20.7	28.7	32.9	35.4	47	46.7
Real estate	24.8	24.1	26.9	19.6	26.7	24.3	27.7	15.4	19.4	26.7
Leasing and business services		75.4	58.8	71.8	44.1	45.9	38.1	40.6	42.3	37.4
Sciences research	51.6	61.2	53.4	64.2					38.5	39.4
Resident services	50.2	50.0	50.9	57.8						
Education	28.2	45.9	38.7	38.1					25.6	43.7
Health, social security, and social welfare	57.2	66.6	50.1	68.9					37.4	38.2
Culture, sports, and recreation	53.7	52.0	53.4	56.5					49.4	43.9
Public administration and social organizations	52.2	54.9	49.2	46.7					39.2	36.7

long-term intermediate inputs change settings. Specifically, the rate of change of medium- and long-term intermediate inputs set the following:

1. In the agricultural sector, intermediate inputs will continue to increase, this is because with the changes in the transfer of labor, mechanization and scale production, agricultural machinery, fertilizers, and other inputs for agriculture will continue to increase. Japan and the United States also show a similar trend.
2. Intermediate inputs of energy and resources sectors of the various departments will be slightly lower. With the gradual increase in the price of energy resources, the use of resources will be more economical, so the intermediate input rate calculated in constant prices will be slightly lower.
3. In the intermediate input rate of labor-intensive sectors, there will be some decline. This will mainly be due to the twelfth five year plan. Both medium- and long-term wage levels will be significantly improved, and some proportion of the labor remuneration will increase accordingly.
4. In the intermediate inputs, the production services sector's proportion has increased, mainly on account of the utilization rate of the industrial sector. The service sector has increased due to a certain degree of development experience.

Notes

[1] David Roland-Host, Li Shantong, Theory and implementation of policy modeling techniques CGE Model, Beijing: Tsinghua University Press, 2009.
[2] Gao Mengtao, Bi Lanlan, Shi Huili. Permanent income and farmer's saving: evidence from the micro-panel data rural China. The Journal of Quantitative & Technical Economics, 2008. p.40–51.

Appendix C: Glossary

Capital formation rate The ratio of gross capital formation to gross domestic product (GDP) by expenditure approach.

Central region Including six provinces of China. They are Shanxi, Anhui, Henan, Hubei, Hunan, and Jiangxi.

Collapse in central region The phenomenon of downfall of the economic development level and speed in the central region.

Contribution rate The ratio of the increment of sub-sector to that of the total amount.

Demand side management Guide users to optimize the way in consuming electricity through economic, technical, and administrative methods and thus improve the power consuming efficiency.

Direct consumption coefficient Known as intermediate input coefficient, refers to the quantity of value of products or services of the *i* industry consumed directly by the per unit total output of the *j* industry.

Eastern region Including 10 provinces (municipalities) of China. They are Beijing, Tianjin, Hebei, Shandong, Shanghai, Jiangsu, Zhejiang, Fujian, Guangdong, and Hainan.

Electricity elasticity The ratio of growth rate of electricity consumption to that of GDP in a country/region.

Electricity intensity Known as the electricity consumption per unit of GDP, the ratio of electricity consumption to GDP in a country/region.

Electrification level The degree of production and residential electricity consumption, expressed as the proportion of electricity in final energy.

Endogenous variable Variable to be interpreted by the model and its value is calculated by the model.

Energy elasticity The ratio of growth rate of energy consumption to that of GDP in a country/region.

Energy intensity Known as the energy consumption per unit of GDP, the ratio of total energy consumption to GDP in a country/region.

Exogenous variable The known variable determined by factors beyond the model and its value is set artificially.

Final consumption rate The ratio of final consumption expenditures to GDP by expenditure approach.

Final energy consumption The amount of production and residential consumption of primary energy in a given period, with the energy lost in processing and transformation being subtracted.

Gross domestic product (GDP) The final production output of one country's all permanent units in a given period calculated at market price, a key indicator of one country's economy scale.

Gross national income (GNI) The final outcome of primary income distribution of one country's (region) all permanent units in a given period.

Heavy industry Industries in national economy that produce production materials, including energy, chemical industry, metallurgy, automobile, machinery manufacturing, building materials manufacturing.

Industrialization A development process in which the industrial activity gradually obtains dominant position in one country's national economy.

Labor-intensive industry Industries that rely on huge amounts of labor force and on fewer technologies and equipments to undertake production.

Market economy A kind of economic system, under which the production, allocation, and consumption of products are directed by market price and configured by the market instead of being planned and specified by the government.

Negative correlation If a change in a variable is able to cause the opposite change in another variable then the relationship between the two variables is called a negative correlation.

National electricity consumption Refers to the sum of electricity consumption in all industrial production and in residential within a certain period of time. It is composed of primary industry electricity consumption, secondary industry electricity consumption (including power grid line losses), tertiary industry electricity consumption, and urban and rural residential electricity consumption. The first three make up national industrial electricity consumption. Industrial electricity consumption is productive consumption—electric power can drive the operation of machinery and directly produce output value, add value, etc. Residential electricity consumption is a consumer consumption of electricity and essentially does not directly produce any output value.

Non-agricultural industries Including secondary industries and tertiary industries.

Northwestern region Including three provinces of China. They are Liaoning, Jilin, and Heilongjiang.

Electricity consumption per capita The ratio of total electricity consumption to population in one country.

GDP per capita The ratio of GDP to total population in a country/region.

GNI per capita The ratio of GNI to the population in a country/region.

Residential electricity consumption per capita The ratio of total residential electricity consumption to population in one country.

Positive correlation If an increase in a variable is able to cause an increase in another variable, or a decrease in a variable is able to cause a decrease in another variable, namely a concurrent change in variables, then there exists a positive correlation between the two variables.

Planned economy A kind of economic system, under which the production, allocation, and consumption of products are planned and specified in advance by the government.

Primary energy Unchanged and untransformed energies from nature, including raw coal, crude oil, natural gas, and primary electricity (hydropower, nuclear power, wind power, solar power, etc.)

Primary energy consumption The sum of production and residential consumption of primary energy in a given period.

Primary product Unprocessed or slightly processed products.

Resource-intensive industry Industries that can undertake production by consuming large amounts of land, forests, mineral resources, and other natural resources.

Saturation point Generally, it refers to the highest level in development. In this book, it refers to the value when the electricity consumption, energy consumption or product output become stable.

Scenario analysis Assume in advance that several scenarios may occur in the developing process, then set the related factors and use mathematical methods to anticipate and analyze the development under various scenarios.

Technology-intensive industry Industries that need complex and advanced technologies to undertake production.

The twelfth Five-Year Plan The twelfth Five-Year Plan that the Chinese government has made for the national economic and social development, starting from 2011 to 2015.

The troika of the national economy Investment, consumption, and export.

Three demands of the national economy Investment, consumption, and net export.

Three industries Include primary industry, secondary industry, and tertiary industry.

Urbanization The process of rural population becoming urban population.

Urbanization rate The ratio of urban population to the total population in one country.

Western region Including 12 provinces of China (autonomous regions and municipalities). They are Inner Mongolia, Sichuan, Chongqing, Shaanxi, Gansu, Qinghai, Ningxia, Xinjiang, Tibet, Guangxi, Guizhou, and Yunnan.

Printed in the United States
By Bookmasters